X-ray Scattering from Semiconductors

2nd Edition

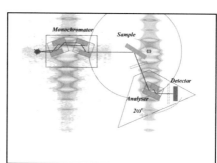

X-ray Scattering from Semiconductors

2nd Edition

Paul F Fewster
PANalytical Research Centre, UK

Imperial College Press

Published by

Imperial College Press
57 Shelton Street
Covent Garden
London WC2H 9HE

Distributed by

World Scientific Publishing Co. Pte. Ltd.
5 Toh Tuck Link, Singapore 596224
USA office: Suite 202, 1060 Main Street, River Edge, NJ 07661
UK office: 57 Shelton Street, Covent Garden, London WC2H 9HE

British Library Cataloguing-in-Publication Data
A catalogue record for this book is available from the British Library.

X-RAY SCATTERING FROM SEMICONDUCTORS (2nd Edition)
Copyright © 2003 by Imperial College Press

All rights reserved. This book, or parts thereof, may not be reproduced in any form or by any means, electronic or mechanical, including photocopying, recording or any information storage and retrieval system now known or to be invented, without written permission from the Publisher.

For photocopying of material in this volume, please pay a copying fee through the Copyright Clearance Center, Inc., 222 Rosewood Drive, Danvers, MA 01923, USA. In this case permission to photocopy is not required from the publisher.

ISBN 1-86094-360-8

Printed in Singapore by World Scientific Printers (S) Pte Ltd

Dedication

To Penelope, Thomas and Anna.

Preface

This book indicates the X-ray scattering methods available to analyse a large range of materials. The emphasis is on the evaluation of the structural properties that influence the physical properties and should be of interest to those who wish to understand more about their material. Semiconductors range in quality from the most perfect crystals to amorphous materials, all of which can be analysed by X-ray scattering, so the methods are common for a large range of other materials.

This book covers the basic structural characteristics of materials, the theory of X-ray scattering for analysing material properties, the principles of the instrumentation, including recent developments, and numerous examples of analyses. Considerable sections of the theoretical development, principles of the instrumentation and examples have not been previously published.

This second edition differs from the first in including a comprehensive subject index, corrections to some of the equations (although most were rather obvious) and changes to some of the figures. This is now closer to my original intentions.

Contents

Chapter 1
An Introduction to Semiconductor Materials

1.1.	General outline	
1.2.	Semiconductors	2
1.3.	Method	6
1.4.	Properties of X-rays	8
1.5.	Instrumentation	11
1.6.	Sample definition	12
References		21

Chapter 2
An Introduction to X-Ray Scattering

2.1.	The interaction of X-ray photons with the sample	23
2.2.	The nature of the scattered X-ray photon with no energy loss	25
2.3.	The near exact theoretical description of scattering	33
	2.3.1. The condition of a single wave generated in a crystal	43
	2.3.2. The condition of two waves generated in a crystal	45
	2.3.3. A further discussion on the deviation parameter β_H	53
2.4.	A scattering theory to accommodate real crystals	56
2.5.	Scattering theory for structures with defects	60
2.6.	Scattering theory of reciprocal space maps	64
2.7.	Approximate theory: The kinematical approach	68
	2.7.1. Comparison between dynamical and kinematical models of diffraction	68
	2.7.2. The important derivations of the kinematical theory	70
	2.7.3. Lateral dimension analysis	74
	2.7.4. Scattering by defects: Diffuse scattering	77
2.8.	Optical theory applied to reflectometry	82
	2.8.1. Some general conclusions from this analysis	88
	2.8.2. Imperfect interfaces	90
2.9.	In-plane scattering	96

2.10.	Transmission geometry	99
2.11.	General conclusions	102
References		103

Chapter 3
Equipment for Measuring Diffraction Patterns

3.1.	General considerations	105
3.2.	Basics of the resolution function	108
3.3.	X-ray source	111
3.4.	X-ray detectors	113
	3.4.1. The proportional detector	114
	3.4.2. The scintillation detector	117
	3.4.3. The solid state detector	118
	3.4.4. Position sensitive detectors	119
3.5.	Incident beam conditioning with passive components	121
	3.5.1. Incident beam slits: Fixed arrangement	121
	3.5.2. Incident beam slits: Variable arrangement	124
	3.5.3. Parallel Plate Collimators	126
	3.5.4. General considerations of slits	127
3.6.	Incident beam conditioning with active components	128
	3.6.1. Incident beam filters	128
	3.6.2. Incident beam single crystal conditioners	129
	3.6.2.1. Single crystal groove conditioners	130
	3.6.3. Multiple crystal monochromators	132
	3.6.4. Multilayer beam conditioners	136
	3.6.5. Beam pipes	138
3.7	Diffractometer options: Combinations with scattered beam analysers	140
	3.7.1. Single slit incident and scattered beam diffractometers	141
	3.7.1.1. Applications in reflectometry	143
	3.7.2. Enhancements to the single slit incident and scattered beam diffractometers	145
	3.7.3. Double slit incident and parallel plate collimator scattered beam diffractometers	146
	3.7.3.1. Enhanced double-slit incident and parallel-plate collimator scattered beam diffractometers	147

		3.7.3.2. Applications for low-resolution in-plane scattering	147
	3.7.4.	Diffractometers using variable slit combinations	150
		3.7.4.1. Applications in reflectometry	150
3.8.	Scattered beam analysers with active components		151
	3.8.1.	The double crystal diffractometer	151
		3.8.1.1. Alignment of high resolution diffractometers	152
		3.8.1.2. Applications of the double crystal diffractmeter	154
	3.8.2.	The triple crystal diffractometer	154
		3.8.2.1. Applications of the triple axis diffractometer	155
	3.8.3.	The multiple crystal diffractometer	155
		3.8.3.1. General considerations of data collection	160
		3.8.3.2. Alignment of multiple crystal diffractometers	161
		3.8.3.3. Three dimensional reciprocal space mapping	164
		3.8.3.4. Applications of multiple crystal diffractometry	166
		3.8.3.5. In-plane scattering in very high resolution	167
3.9.	General conclusions		169
References			170

Chapter 4
A Practical Guide to the Evaluation of Structural Parameters

4.1.	General considerations		171
4.2.	General principles		172
4.3.	Analysis of bulk semiconductor materials		173
	4.3.1. Orientation		174
	4.3.1.1. Surface orientation – the Laue method		174
	4.3.1.2. Determining the orientation by diffractometry		177
		4.3.1.2.2. Monochromator and open detector method	177
		4.3.1.2.3. Multiple crystal diffractometer method	179
	4.3.1.3. Determining polar directions		180
	4.3.2. Revealing the mosaic structure in a bulk sample		182
	4.3.2.1. Mosaic samples with large tilts		182

	4.3.2.2. High resolution scanning methods (Lang method)	184
	4.3.2.3. Multiple crystal methods for revealing mosaic blocks	187
4.3.3.	Characterising the surface quality	192
4.3.4.	Measuring the absolute interatomic spacing in semiconductor materials	195
4.3.5.	Measuring the curvature of crystalline and non-crystalline substrates	197
4.4. Analysis of nearly perfect semiconductor multi-layer structures		200
4.4.1.	The first assumption and very approximate method in determining composition	200
4.4.2.	The determination of thickness	205
	4.4.2.1. Determining the thickness from the fringes close to main scattering peaks	205
	4.4.2.2. Determining the thickness from the fringes in the reflectometry profile	208
4.4.3.	The simulation of rocking curves to obtain composition and thickness	209
	4.4.3.1. Example of an analysis of a nearly perfect structure	210
	4.4.3.2. Direct analysis from peak separation and fringe separations	212
	4.4.3.3. Simulation using an iterative adjustment of the model	212
	4.4.3.3.1. Linking parameters to cope with complex multi-layer structures	213
	4.4.3.4. Automatic fitting of the data by simulation	215
	4.4.3.5. Data collection with the 2-crystal 4-reflection monochromator and 3-reflection analyser	219
	4.4.3.6. Reciprocal space map to analyse the imperfections in samples	220
	4.4.3.7. Taking account of tilts in rocking curve analyses	224
	4.4.3.8. Modelling the extent of the interface disruption in relaxed structures	226

	4.4.3.9.	Detailed analysis to reveal alloy segregation and the full structure of a multi-layer	227
	4.4.4.	Analysis of periodic multi-layer structures	229
		4.4.4.1. The analysis using direct interpretation of the scattering pattern	229
		4.4.4.2. The analysis using basic kinematical theory	231
		4.4.4.3. Analysis of periodic multi-layers with dynamical theory	235
		4.4.4.4. Analysis of periodic structures with reflectometry	238
		4.4.4.5. Analysis of a nearly perfect epitaxial periodic multi-layer	239
		4.4.4.5.1. Analysis based on the kinematical approach	240
		4.4.4.5.2. Analysis based on the optical theory with reflectometry	245
		4.4.4.5.3. Analysis based on the dynamical theory simulation	247
4.5.	Analysis of mosaic structures (textured epitaxy)		248
4.6.	Analysis of partially relaxed multi-layer structures (textured epitaxy)		249
	4.6.1.	Measuring the state of strain in partially relaxed thin layers	251
	4.6.2.	Obtaining the composition in partially relaxed thin layers	253
	4.6.3.	The measurement of the degree of relaxation and mismatch in thin layers	255
	4.6.4.	The determination of relaxation and composition with various methods	256
		4.6.4.1. Determination by reciprocal space maps on an absolute scale	256
		4.6.4.2. Determination by using a series of rocking curves and analyser scans	258
		4.6.4.3. Determination by reciprocal space maps on a relative scale	260
		4.6.4.4. Determination by rocking curves alone	261
		4.6.4.5. Revealing dislocations and defects by topography	263

	4.6.4.6. Simulating structures with defects	264
4.7.	Analysis of laterally inhomogeneous multi-layers (textured polycrystalline)	266
	4.7.1. Direct analysis of laterally inhomogeneous multi-layers	266
	4.7.2. Simulation of laterally inhomogeneous multi-layers	270
	4.7.3. Lateral inhomogeneities without large misfits	272
	4.7.3.1. Analysing epitaxial layers with very small twinned regions	273
	4.7.3.2. Analysing twin components larger than 5 microns	274
4.8.	Analysis of textured polycrystalline semiconductors	275
4.9.	Analysis of nearly perfect polycrystalline materials	277
	4.9.1. Measurement of thickness of CrO_x on glass	278
	4.9.2. Analysis of very weak scattering	280
4.10.	Concluding remarks	282
References		282

Appendix 1
General Crystallographic Relations

A.1.	Introduction	285
A.2.	Interplanar spacings	285
A.3.	Stereographic projections	287

Subject Index — 295

CHAPTER 1

AN INTRODUCTION TO SEMICONDUCTOR MATERIALS

1.1. General outline

Semiconductor materials exist in many structural forms and therefore require a large range of experimental techniques for their analysis. The intention of this book is to allow the reader to obtain a good working knowledge of X-ray diffraction techniques so that he or she is fully aware of the possibilities, assumptions and limitations with this form of structural analysis.

The molecular structure of semiconductors is in general well known. Most materials of interest have been manufactured in some way, therefore an approximate knowledge of the elements and layer thicknesses and sequence is assumed and is the starting point for many of the approaches used. Most samples of interest, however, are not of a simple molecular form but are composite structures, commonly consisting of multiple thin layers with different compositional phases. There are many important structural parameters that can modify semiconductor device performance. These parameters include phase composition, micro-structural or layer dimensions and imperfections, etc. A description of how the properties are categorised and material form is given, since this largely determines the X-ray scattering experiment for the analysis, Fewster (1996).

The theory of X-ray scattering is presented from a physical basis and therefore naturally starts with dynamical theory and its extensions before describing the more approximate kinematical theory. The theories are largely considered at the single photon level. In reality the experiment collects many photons that are divergent and occupy a range of energies.

These effects influence the experimental results and are covered in the description of the instruments. In general the scattering is three-dimensional because the sample is three-dimensional and this is born in mind throughout. The mapping of the scattering in three-dimensions is the most general experiment and all other approaches are obvious projections. The assumptions on moving to two-dimensional reciprocal space mapping and ultimately one-dimensional X-ray scattering, e.g. rocking curves will then appear more obvious and understandable. The assumptions associated with interpreting data collected in various ways will be discussed. This will then allow the reader to understand the subject conceptually and extend the techniques to his or her particular problem.

The concentration is on near-perfect semiconductor materials defined by the valence in the Periodic Table, that is group IV, III-V, II-VI, etc., since the available information is large as well as being commercially very important. The concentration on near-perfect materials is to contain the range of techniques and available information. The techniques will be considered in general terms so as not to limit the book to specific materials.

Interpretation requires understanding of the instrument aberrations, the assumptions in the methods and presumed details concerning the sample. Aspects of quality control with X-ray diffraction methods based on comparative measurements requires an understanding of the sensitivity of the conditions to the parameters of interest, these aspects will also be covered.

1.2. Semiconductors

Semiconductors range from the most perfect crystals available to amorphous materials. The sample dimensions can range from 12 inch diameter ingots of bulk Si to layers of partial-atomic coverage layers with nanometre scale lateral dimensions embedded in a multi-layer structure. The molecular structure of most common semiconductors is of a high symmetry extended lattice, for example Si has the space group (this represents the relative relationships between atoms) of $Fd3m$ and the space group for GaAs and InP is $F\bar{4}3m$. Both these space groups are simple face-centred cubic lattices, figure 1.1. Also the degree of complexity in semiconductor samples is increasing with the manufacture of laterally patterned or phase separated structures having dimensions sufficiently small to create "zero" dimensional quantum size effects. The list of structural parameters required to define these materials is increasing, although the basic definitions can be described quite simply, Table 1.1. Since these structures are manufactured the analysis has a good starting point in that an approximate understanding of the structure will exist. The challenge therefore is to determine these parameters more precisely, not only to aid the manufacturing process but also to analyse them for defects and structural quality, etc.

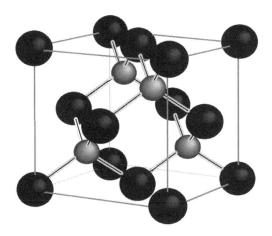

Figure 1.1 The arrangement of atoms in typical III-V semiconductor structures (e.g. GaAs, InP).

All physical properties, be they electronic or optical, rely in some way on the structural properties of the material. As the degree of device sophistication increases the structural tolerances are reduced. Therefore the necessity of accurate determination of the composition, thickness and defects, etc., becomes very important. X-ray analysis techniques are very well developed for obtaining this information and mature enough to recognise its weaknesses. Of course no technique should be used in isolation but compared and complimented with other methods. Most techniques measure something that does not compare directly with any other, this in itself gives valuable information not only on the assumptions in deriving the information but also the sensitivity of the technique.

Amorphous semiconductors are important in the areas of large area electronics, although because of the low free-carrier mobility some of the associated circuitry requires re-crystallisation. This change of structural form improves the mobility and device speed. The sizes of the crystallites, the location of crystallite boundaries with respect to the device active region are all-important parameters. As we increase the degree of perfection the long-range structural order gives rise to well defined electronic band structures. This leads into the possibilities of band structure engineering. The variables that the device engineer has are composition, shape and orientation to create structures with the required optical or electronic responses. Control over the growth of these structural parameters at the required level, sometimes at the atomic level, is not a trivial task. The growth of these structures is dominated by epitaxy either from the liquid phase (LPE), chemical vapour deposition (CVD), molecular beam epitaxy (MBE) or extended forms of these; metal organic CVD (MOCVD), metal organic MBE (MOMBE), etc. These growth methods require very careful control and therefore careful analysis (in-situ and ex-situ) to ensure the structural parameters are those wanted. Also different compositional phases have different interatomic spacings and therefore these must be accommodated either by elastic strains or plastic deformation. Plastic deformation exists in the form of cracks and dislocations, which can act as charge carrier recombination centres and alter the device performance. It is therefore very necessary to have knowledge of the defects in the active region of the device that controls their behaviour. Again very careful structural analysis is required. All these properties can depend strongly on the quality of the substrate material, i.e. its defect density, orientation and surface strains, etc.

It should be clear that these semiconductors cannot be grown by pressing a few buttons and achieving the performance expected. A very good and thorough understanding of the materials and the growth method are required. In-situ analysis methods to monitor the growth are developing but generally the most thorough analysis are performed ex-situ. The in-situ methods generally rely on a detailed understanding from post-growth analysis that of course can be very exhaustive. X-ray diffraction methods are sometimes used in-situ but in general contribute to improving yield by analysing material at various stages in manufacture, help in controlling the growth process and for detailed materials analysis ex-situ.

Table 1.1 Definition of the structural properties of materials.

Type of property	General property	Specific property
Macroscopic	*Shape*	Layer thickness
		Lateral dimensions
	Composition	Structural phase
		Elements present
		Phase extent
	Form	Amorphous
		Polycrystalline
		Single Crystal
	Orientation	General preferred texture
		Layer tilt
	Distortion	Layer strain tensor
		Lattice relaxation
		Warping
	Homogeneity	Between analysed regions
	Interfaces	Interface spreading
	Density	Porosity
		Coverage
Microscopic	*Shape*	Average crystallite size
		Crystallite size distribution
	Composition	Local chemistry
	Orientation	Crystallite tilt distribution
	Distortion	Crystallite inter-strain distribution
		Crystallite intra-strain distribution
		Dislocation strain fields
		Point defects
		Cracks
		Strain from precipitates
	Interface	Roughness laterally
	Homogeneity	Distribution within region of sample studied

The X-ray analysis technique to apply depends on the material quality, the level of detail and precision required. This book will describe all the levels of precision and assumptions made to carry out certain types of analysis. Because the crystalline quality of many semiconductors is very high the diffraction process cannot be treated in a simple way. Most analyses require the application of the dynamical diffraction theory and therefore an understanding of this and the assumptions involved are important and described in Chapter 2. The development of instrumentation for collecting the scattered X-rays has also created new possibilities in analysis that make X-ray methods a very versatile tool in probing the structure of materials.

1.3. Method

In this section a brief description of the accessible information to X-ray diffraction techniques will be given. How and why this following information is possible to extract will become clear in later chapters. Table 1.1 presents the definition of various structural parameters used to define a material. The first subdivision of the structural properties is into macroscopic and microscopic. These are X-ray definitions and can be considered respectively as aspects that define the major features of the diffraction pattern (peak position and intensity) and those that alter the pattern in a more subtle way (peak shape and weak diffuse scattering).

Table 1.2. Definition for structural types.

Structural type	Definition
Nearly perfect epitaxial	A single extended crystal having near perfect registry with the same orientation as the underlayer, which is also nearly perfect.
Textured epitaxial	The layer orientation is close to registry with the underlayer, both normal and parallel to the surface plane. The layer is composed of mosaic blocks.
Textured polycrystalline	Crystallites preferentially orientated normal to the surface, but random in the plane. They have a distribution in sizes.
Nearly perfect polycrystalline	Random orientated crystallites of similar size and shape.
Amorphous extended lattice	Similar strength interatomic bonds but no length scale correlation greater than this.
Random molecules	Essentially amorphous structure with weak interlinking between molecules, possibly giving some ordering.

X-ray diffraction is a very sensitive structural analysis tool and the extent to which detailed information can be obtained depends on the sample itself. Suppose that the sample is poorly defined and contains numerous crystallites, with a distribution of structural phases, sizes, orientations and strains, then separating the various contributions is not trivial. However if certain properties can be determined rather precisely then others can be determined by extending the range of experiments. Clearly therefore, the initial assumptions concerning the sample will define which structural details can be obtained readily. At this stage we should define the sample since these will define the likely information that can be determined by X-ray methods and the type of instrumentation that is applicable, Table 1.2.

Orientation in this context refers to the alignment of low index atomic planes (these are planes separated by distances of about one unit cell spacing) to some other reference, e.g. the surface. These definitions concentrate on laterally extended homogeneity and therefore can be expanded to included patterned structures and random structural variations in the lateral plane by considering them as columns.

Since any structural probe will determine an average of a region or analyse an unrepresentative region of an inhomogeneous sample, it is clear to see that the useful information may be limited to some average parameter and its variation. For structural types that are highly inhomogeneous, e.g. random molecules and textured polycrystalline materials, then X-ray diffraction will average some long-range order, orientation distribution and their variations. This is where it is important to link the physical or chemical property of the material to the structural property, for example is it the macroscopic average or the microscopic details that determines the property of interest? The next question may well be the scale of the variation; is it homogeneous at the micron or nanometre scale? X-ray diffraction averages in several ways, within a coherently diffracting volume and the X-ray beam dimensions on the sample. Controlling the beam divergence can modify the former and the latter can be subdivided by analysing the scattered beam with an area detector as in X-ray topography. The range of X-ray analysis techniques therefore cannot be simply categorised into finite bounds of applicability but depend upon the material, the property of the material of interest, the versatility of the diffractometer, the X-ray wavelength, etc. Understanding the details of diffraction process, the nature of X-rays and assumptions concerning the sample are all-important to making a good and reliable analysis.

Some typical macroscopic and microscopic properties are given in figures 1.2 and 1.3 respectively. The important aspect here is the X-ray probe dimension with respect to the properties. Clearly the probe is not simply defined two dimensionally but also has some depth into the figures, consequently we must be aware how this probe brings all this information together to create a signal which is then interpreted. Having defined some basics concerning the sample we shall consider some basic information about the X-rays used to extract this information.

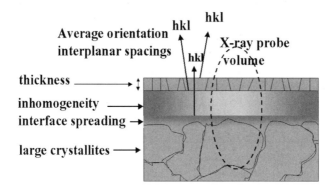

Figure 1.2 The main macroscopic parameters that characterise a layered structure.

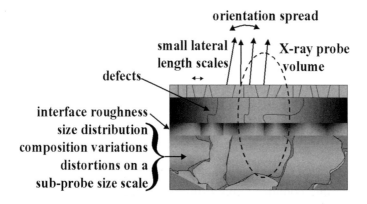

Figure 1.3 The main microscopic parameters that characterise a layered structure.

1.4. Properties of X-rays

X-ray wavelengths compare with the energy transitions of inner electron orbitals in atoms. It is this property that is used to create laboratory monochromatic X-rays. High energy decelerating electrons will also emit X-radiation and this is the reason for the continuum of radiation from laboratory sources and the emission from synchrotron sources. A laboratory source is shown diagrammatically in figure 1.4 with an accompanying spectrum. The radiation from a laboratory source is not very uniformly distributed and in general much of the radiation is not used. However the intense characteristic lines act as a good internal standard and it is these lines that are used in the majority of laboratory experiments.

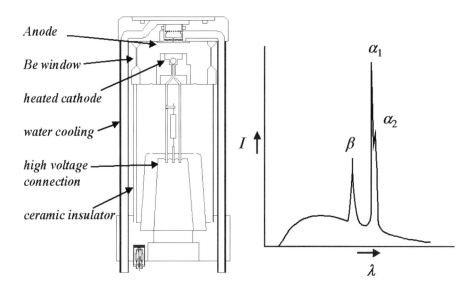

Figure 1.4 The interior of a modern sealed source X-ray tube and the spectral variation with intensity for a typical anode material

Synchrotron radiation sources work on a very different principle. A synchrotron is really a storage ring for electrons, which are contained by magnetic fields to prevent excessive divergence and consequent energy loss. When the electrons are deviated from a straight line using so-called bending magnets, wigglers and undulators, the consequent acceleration towards the centre of the curve creates an energy orbital jump thus producing electromagnetic radiation. If this energy change is large (i.e. high speed electrons and small bending radius from intense magnetic fields) then X-rays can be produced. The X-rays from the synchrotron are emitted tangentially from the radius and concentrated into a narrow cone with the electric field vector predominately confined to the plane of the orbit; i.e. the beam is horizontally polarised. However it is possible to rotate this plane of polarisation but in general this aspect does restrict most experiments to scattering in the vertical plane. This can lead to extra tolerances required for mechanical movements of diffractometers because of the gravitational pull. Another aspect to consider is the wavelength calibration; this has to be done before any experiment since the emission is smooth and there are few reference lines except at absorption edges. Laboratory sources are rather less efficient at producing X-rays. The emerging X-rays are randomly polarised and almost radially symmetric, yet only a small percentage of this divergent source can be used.

Because of the method of injecting electrons into a synchrotron they are arranged in bunches and therefore the consequential X-ray emission will have a time structure. This can prove useful for some experiments especially when only a single bunch is injected, if the X-ray pulse can be synchronised with some dynamic experiment, (Barrington-Leigh and Rosenbaum, 1976, Whatmore, Goddard, Tanner and Clark, 1982). Third generation sources are also increasing the brilliance level. This gives rise to very small source sizes that can create some phase coherence across the whole source leading to observable interference effects when the beam travels along different optical paths (phase contrast topography and tomography, Cloetens et al, 1999). This coherence over the source can also create routes to reconstructing the scattering object (Miao, Charalambous, Kirz and Sayre, 1999). Laboratory X-rays from similar sized sources have very low power output. However there are methods of effectively moving the source close to infinity with crystal optics: the phase front formed from an object containing several different optical paths can then be separated with an analyser crystal (Davis, Gao, Gureyev, Stevenson and Wilkins, 1995). It is clear that the developments and possibilities continue and this is far from a static subject. These developments will then lead to new possibilities in analysis.

The highly directional aspects of the synchrotron generated X-rays leads to very intense sources compared with laboratory sources. However the convenience and improvements in intensity output makes the laboratory sources suitable for most experiments. One of the earliest methods of increasing the intensity in the laboratory was achieved by rapidly rotating the anode (rotating anode source) to distribute the heat. This has lead to increases in intensities by almost an order of magnitude for 15kW sources, although 60kW sources are available. However there are many other ways of improving the intensity and whatever method is used it has to be related to the problem to be solved, since the intensity output should be qualified with flux, divergence, wavelength distribution, etc.

Since X-rays are primarily generated from inner atom core transitions the photon wavelengths are in the region of 0.1nm, which is of the order of the interatomic spacings in materials. Bragg's equation (derived in Chapter 2) indicates that the difference in the scattering angle of two interatomic spacings of 0.14 and 0.15 nm determined with a 0.15 nm X-ray wavelength is ~2.4^0. As will be seen later the peak widths of diffraction maxima can be located within about 0.0002^0. This gives X-rays the high strain sensitivity at the part per million level and sensitivity to atomic scale spatial resolutions.

1.5. Instrumentation

There have been considerable developments in new instrumentation. The power of laboratory X-ray sources have increased and various focusing mirrors and X-ray lenses can recover the divergence of laboratory X-ray sources with considerable intensity enhancements. The degree of sophistication is increasing with the various components recognising each other (i.e. exchanging the X-ray tube will be recognised by the system, thus limiting the power delivered, etc.). Computer automation has made considerable improvements in time and freed user involvement and this will continue. This has considerably helped in the thinking to doing ratio.

The mechanical stability has also improved with optical encoding on the axes, allowing fast movement to very high precision. Interchangeable components (monochromators, X-ray mirrors, slits, etc.) increase the versatility of diffractometers and can be pre-aligned so that several very different experiments can be performed on one instrument with a simple change. The experiment can now be fitted to the sample and property of interest instead of the former more established approach of having an array of instruments for each experimental technique.

The choice of instrumental configuration and its consequential influence on the information required from the sample will be covered in Chapter 3.

1.6. Sample definition

The various sample types have been described in section 1.2 and 1.3, but here the description will be defined more closely to that required for X-ray diffraction. These definitions also indicate the information of importance in analysis, Table 1.3. Basically any structure will be an arrangement of atoms. However a crystal is defined as "any structure having essentially a discrete diffraction pattern." This is the accepted definition (Acta Crystallographica **A48** 928, 1992). To have a diffraction pattern that is observable with X-rays in the simplest case requires some form of periodicity or repeat unit cell.

A semiconductor, for example GaAs, Si, GaN, consists of an extended periodic array and would fit into the above categories of perfect epitaxy, textured epitaxy and possibly textured polycrystalline in thin layer form. Although material with no dislocations (missing lines of atoms) can be grown, most do have dislocations threading through them. The generation of convection currents during growth can create mosaic blocks (crystallites surrounded by defects) that can be tilted with respect to each other. These are all fairly typical features found in bulk material and thin films. One of the most fundamental problems in thin films is that the atomic spacing of the layer differs from that of the underlying material. This will cause either elastic distortion or, if the internal stress exceeds that which can be accommodated by elastic strains, plastic deformation occurs and misfit dislocations are generated. Misfit dislocations can be formed from the high stress levels and imperfections at the growing surface nucleating dislocation loops that glide to the interface or by turning a threading dislocation to lie in the interface plane. Knowledge of the state of strain, the number of defects, etc., can be very important for device performance and X-ray diffraction methods are very sensitive to these effects.

Figure 1.5 gives a three dimensional view of the structural properties of a thin film. Basically we have a unit cell repeat that can vary laterally and in depth, having parameters a, b, c, α, β and γ. Within this there are relative rotations between regions and layers, defects (dislocations and point defects (atomic site errors, e.g. interstitials, vacancies and impurity atoms)). These features all influence the diffraction pattern of X-rays. The creation of the scattering pattern from X-rays is one thing but to interpret the features is quite another and a reasonable understanding of the sample in question is necessary. The important aspects will now be considered here. The very high strain sensitivity can allow measurement of $x < 1\%$ (absolute) composition variations in $Al_xGa_{1-x}As$ alloys, $<0.1\%$ (absolute) composition variations in $In_xGa_{1-x}As$, etc., for peak shifts of $0.001°$ or 3.6 seconds of arc. However to achieve this the strain has to be related to the composition using some assumptions. As discussed above a thin layer grown epitaxially on a substrate will distort either elastically or plastically. In both cases we need to determine the unit cell parameters of the layer of interest and calculate how this would change if it were free standing. Clearly we have to include the influence of elastic parameters.

Table 1.3. A broad overview of the structural parameters that characterise various material types. Those parameters that have meaning in the various materials are given with filled diamonds, those that could have meaning are given by open squares.

	Thickness	Composition	Relaxation	Distortion	Crystallite size	Orientation	Defects
Perfect Epitaxy	♦	♦				♦	
Nearly perfect epitaxy	♦	♦	□	□	□	♦	♦
Textured epitaxy	♦	♦	♦	♦	♦	♦	♦
Textured polycrystalline	♦	♦	□		♦	♦	□
Perfect polycrystalline	♦	♦			♦	♦	□
Amorphous layers	♦	♦					

The arrangement of atoms in silicon is similar to that given in figure 1.1, except that all the atoms are identical. If we try and compress the structure along the bonds, [111] type directions then it will be much more difficult to do so than along a [100] direction where we would distort angles, for example. So although the structure is of high symmetry, its elastic properties are very anisotropic. Compression along one direction will necessitate an expansion in another. This can be characterised by examining the relationship between stress and strain. Initially we can suppose that the strain is elastic until the internal stress is too large and plastic deformation occurs. The plastic deformation will occur as cracks or dislocations, however the strain parallel to any interface will be related to the degree of alignment of the atoms in a layer with that underneath. Hooke's law gives the relationship of stress to strain, but because we are considering an anisotropic medium we have to generalise the problem and the elastic stiffness to fourth rank tensors, Nye (1985). This is all rather unwieldy and can be simplified to a 6 × 6 matrix when equivalent coefficients are considered

$$\begin{pmatrix} \sigma_{xx} \\ \sigma_{yy} \\ \sigma_{zz} \\ \sigma_{yz} \\ \sigma_{xz} \\ \sigma_{xy} \end{pmatrix} = \begin{pmatrix} c_{11} c_{12} c_{13} c_{14} c_{15} c_{16} \\ c_{21} c_{22} c_{23} c_{24} c_{25} c_{26} \\ c_{31} c_{32} c_{33} c_{34} c_{35} c_{36} \\ c_{41} c_{42} c_{43} c_{44} c_{45} c_{46} \\ c_{51} c_{52} c_{53} c_{54} c_{55} c_{56} \\ c_{61} c_{62} c_{63} c_{64} c_{65} c_{66} \end{pmatrix} \begin{pmatrix} \varepsilon_{xx} \\ \varepsilon_{yy} \\ \varepsilon_{zz} \\ \varepsilon_{yz} \\ \varepsilon_{xz} \\ \varepsilon_{xy} \end{pmatrix} \qquad 1.1$$

c_{ij} are the stiffness coefficients, where c_{11} is the stiffness along the a axis and c_{46}, etc., give rise to shear. σ_{ij} and ε_{ij} represent the stresses and strains along various directions. This is the general case for triclinic structures. For higher symmetry many of these coefficients become zero and some become equivalent.

Suppose we consider the growth along the z <001> direction then the layer will be constrained in the plane of the interface and the stress will be zero normal to this direction, i.e. the top surface is unconstrained then

$$\sigma_{zz} = 0 = c_{31}\varepsilon_{xx} + c_{32}\varepsilon_{yy} + c_{33}\varepsilon_{zz} + c_{34}\varepsilon_{yz} + c_{35}\varepsilon_{xz} + c_{36}\varepsilon_{xy} \qquad 1.2$$

Now for a cubic system $c_{31} = c_{32} = c_{13} = c_{23} = c_{12} = c_{21}$, $c_{22} = c_{33} = c_{11}$ and $c_{34} = c_{35} = c_{36} = 0$, therefore

$$\varepsilon_{zz} = -\frac{c_{12}}{c_{11}}\{\varepsilon_{xx} + \varepsilon_{yy}\} \qquad 1.3$$

Hornstra and Bartels (1978) have solved the condition for the general case of cubic systems, giving the calculation procedure and some examples for the GaAs phase. Any solutions of the cubic system will just include combinations of the coefficients c_{11}, c_{12} and c_{44}.

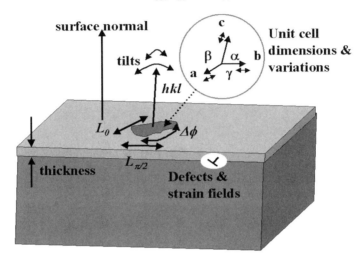

Figure 1.5 The range of structural parameters accessible to X-ray methods for imperfect (real) samples.

As the degree of anisotropy decreases a Poisson ratio can be used (in fact a Poisson ratio can be used for any system except that it varied with direction). The equivalent Poisson ratio for the above example with the <001> direction normal to the surface direction is therefore

$$v = \frac{\varepsilon_\perp}{\varepsilon_\perp - 2\varepsilon_{//}} = \frac{c_{12}}{c_{11} + c_{12}} \qquad 1.4$$

where

$$\varepsilon_{//} = \frac{1}{2}\{\varepsilon_{xx} + \varepsilon_{yy}\} = \frac{1}{2}\left\{\frac{_L d_x - _{L0} d_x}{_{L0} d_x} + \frac{_L d_y - _{L0} d_y}{_{L0} d_y}\right\} \qquad 1.5$$

and

$$\varepsilon_\perp = \varepsilon_{zz} = \frac{_L d_z - _{L0} d_z}{_{L0} d_z} \qquad 1.6$$

where $_L d_x$, etc., are the actual atomic plane spacings along x and $_{L0} d_x$ are the unstrained or free standing atomic plane spacings along x, etc. Since we wish to know the strain normal to the interface due to the strain parallel to the interface we can write the general case (remembering that ν is a function of direction)

$$\varepsilon_{zz} = \frac{-\nu}{1-\nu}(\varepsilon_{xx} + \varepsilon_{yy}) \qquad 1.7$$

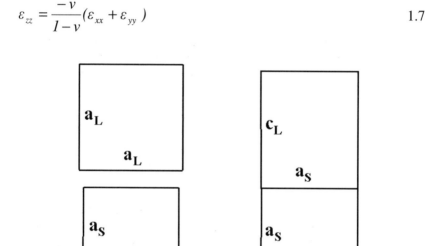

Figure 1.6 The undistorted (before deposition) and a distorted (after deposition) unit cell for a simple cubic layer on a cubic substrate; both are orientated along a cubic edge direction.

If constant volume is maintained during distortion then Poissons's ratio is simply 0.5, however most materials, especially the structural form of typical III-V semiconductors this is much closer to 0.3. Clearly as the molecular form becomes more complicated and the bond directions more random this anisotropy will decrease. Additional shear actions come into effect along directions of lower symmetry, however these may be of less concern in a homogeneous thin layer of large lateral dimensions.

Let us firstly consider a cubic (001) GaAs substrate with a thin film of cubic AlAs on top. Both structures have the same space group, the same arrangement of atoms, but slightly different lattice parameters and elastic parameters. If a thin layer (~0.2 µm) is deposited then the atoms will align with those of the substrate and the structure will appear as a continuous lattice with an abrupt change in lattice parameter and composition at the substrate interface. The alignment of the atoms in the interface plane will define the lattice parameter of the layer in the interface plane and through the appropriate Poisson ratio or elastic stiffness combinations will define the lattice parameter normal to the interface plane, figure.1.6.

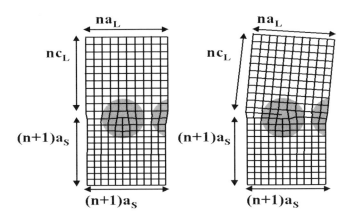

Figure 1.7 The problems that occur when the elastic parameters are incapable of accommodating the distortions necessary for perfect epitaxy.

As the thickness of the layer increases the layer will become progressively more reluctant to distort. Eventually the elastic limit will be reached and only partial registry will exist. When this situation arises there will be more rows of atoms in the substrate than in the layer (the lattice parameter of GaAs is less than that of AlAs). The average lattice parameter in the plane of the interface will therefore differ and some distortion will extend into the layer and substrate with possible tilting between the two, figure.1.7. Clearly the structure becomes quite complex even for this simple system. For perfect epitaxy we are trying to match the layer interatomic spacings to that of the layer or substrate below, $_s d_x$, etc., however if the layer has partially relaxed back to its strain-free state then we can rewrite equation 1.5 as

$$\varepsilon_{//} = \frac{1}{2}\{\varepsilon_{xx} + \varepsilon_{yy}\} = \frac{1}{2}\left\{\frac{_s d_x - _{L0} d_x}{_{L0} d_x}[1-R_x] + \frac{_s d_y - _{L0} d_y}{_{L0} d_y}[1-R_y]\right\} \quad 1.8$$

where R_x is the relaxation in the misfit along x, etc. The relaxation is then zero for perfect matching and unity when the layer relaxes to its unconstrained shape. Substituting equation 1.8 into 1.7 will therefore give the perpendicular strain from knowledge of the original lattice parameters of the component layers and the degree of relaxation in two orthogonal directions.

$$\varepsilon_\perp = \frac{-v}{1-v}\left\{\frac{_s d_x - _{L0} d_x}{_{L0} d_x}[1-R_x] + \frac{_s d_y - _{L0} d_y}{_{L0} d_y}[1-R_y]\right\} \quad 1.9$$

When the unit cells of the two materials differ significantly then the registry of the atoms becomes very complex. Consider for example (0001) GaN on (0001) sapphire, both are hexagonal structures but the lattice parameters differ quite considerably. The atom arrangement of both materials is given in figure.1.8. The best atomic match appears when the two lattices are aligned along different directions, i.e. the x-direction is rotated through 90^0 with respect to the other. However we can see that this gives approximate alignment of the Al in the sapphire to the Ga in the GaN. The mismatch is so large however that the GaN is very heavily relaxed towards its unstrained state and will be full of defects associated with this poor match.

Pashley (1956) has given a very full account of the possibilities in epitaxy and the accounts of the early theories. A full all-encompassing theory explaining the nucleation and orientation dependence is still elusive, but there are some general guidelines that can be given. Generally an orientation dependence occurs when the mismatch (the fractional difference in the lattice plane spacing in the plane of the interface) between the overlayer and underlayer is less than ~14%. Theoretical models and experimental evidence on a wide range of systems support this. The orientation depends on the relative alignment of atoms in the overlayer and underlayer and is not governed by integer relationships of atomic plane spacings of the two lattices. The thickness of the layer influences the extent to which the elastic distortion can be accommodated, the greater the misfit the thinner the layer should be to maintain good epitaxy. Once epitaxial growth is established in the fabrication of a structure then a full understanding of these nucleation processes may seem irrelevant, however new structures are emerging that make use of some of the nucleation properties of certain materials.

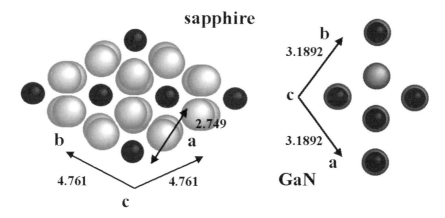

Figure 1.8 The basal plane view of GaN and sapphire; this indicates the rotation necessary to accommodate the alignment of atoms for epitaxy. Al and Ga are coloured black.

We can consider growth to occur in three principle ways. The first is a two-dimensional mechanism, i.e. the layer is built up atomic layer by atomic layer, and this relies on the atoms migrating across the surface and preferring to locate at atomic layer steps. A complete atomic layer coverage will create a very smooth surface, whereas for intermediate coverage the surface is atomically rough. This oscillation in smooth and rough surfaces explains the oscillating specular reflectivity observed in Reflection High Energy Electron Diffraction (RHEED) during growth by Molecular Beam Epitaxy (MBE), Neave, Joyce, Dobson and Norton (1983). When the surface does not "wet" easily with the deposited atoms of the overlayer, the growth can occur in distinct islands that gradually enlarge and eventually coalesce. This is termed three-dimensional growth and can lead to mosaic or columnar growth with defects concentrated at the boundaries. Another interesting growth mode is a mixture of both three- and two-dimensional growth first described by Stranski and Krastanow (1938). This mechanism is characterised by the initial formation of a wetting layer (two-dimensional growth) that is very thin (no more than a few atomic layers) and the subsequent growth of islands. Examples of these mechanisms can be seen in semiconductor materials. InGaAs deposited on GaAs at low In compositions, <10%, will grow two-dimensionally up to about 70nm before misfit dislocations are formed at the interface, whereas InAs will grow by the Stranski-Krastanow mechanism. These differing growth modes may appear troublesome but can be used to advantage in creating structures defined in all three dimensions by optimising the growth method. These can have very special properties and offer another challenge to analytical methods.

Determining the growth mode can only be accomplished with precise surface diffusion data and bond strengths, etc. It is then possible to construct a surface by modelling the whole process and extracting a statistical significance. These approaches have been very successful at predicting some of the general observations of surface topography, Itoh, Bell, Avery, Jones, Joyce and Vvedensky (1998).

Predicting the situation when defects form at the interface between two materials, i.e. when the elastic limit has been exceeded, has been the subject of many studies. This of course is a very important parameter because defects in general are detrimental to semiconductor devices and knowledge of the bounds of lattice parameter misfit and thickness define whether a device is possible to fabricate. Hull and Bean (1992) have reviewed the mechanisms of dislocation generation and propagation and discussed the definitions and derivation of the "critical" thickness defining their onset. Of course there are many experimental studies that have questioned the theoretically derived values. Dunstan, Kidd, Howard and Dixon (1991) have taken a very pragmatic approach to the evaluation of critical thickness and compared the residual strain as a function of thickness. The resulting curve is remarkably predictable for a large range of material systems and offers a very quick procedure for predicting the onset of relaxation.

If the mismatch is large or the substrate is amorphous then the orientation dependence of the layer can be governed by very different criteria and the layer can become essentially textured polycrystalline or even random polycrystalline. Knowledge of the likely form of these materials will define the type of experiment necessary to obtain detailed structural information.

Chapter 2 will describe the theoretical basis of scattering from various structures typically encountered in the field of semiconductor physics. These materials represent the case when a large amount of information is available, although the techniques are applicable to any material and not specific to semiconductors.

References

Barrington-Leigh, J and Rosenbaum, C (1976) Ann. Rev. Biophysics and Bioengineering **5** 239.
Cloetens, P, Ludwig, W, Baruchel, J, Guigay, J-P, Pernot-Rejmankova, Salome-Pateyron, M, Schlenker, M, Buffiere, J-Y, Maire, E and Peix, G (1999) J Phys. D: Appl. Phys. **32** A145.
Dunstan, D J, Kidd P, Howard, L K and Dixon, R H (1991) Appl. Phys. Lett. **59** 3390.
Fewster, P F (1996) Rep. Prog. Phys. **59** 1339.
Hornstra, J and Bartels, W J (1978) J Cryst. Growth **44** 513.
Hull, R and Bean, J C (1992) Critical Reviews in Solid State and Materials Sciences **17** 507.
Itoh, M, Bell, G R, Avery, A R, Jones, T S, Joyce, B A and Vvedensky, D D (1998) Phys. Rev. Lett. **81** 633.
Miao J, Charalambous, P, Kirz J and Sayre D, (1999) Nature **400** 342.
Neave, J H, Joyce, B A, Dobson, P J and Norton, A (1983) Appl. Phys. Lett. **A31** 1.
Nye, J F (1985) Physical Properties of Crystals – Their representation by Tensors and Matrices Oxford Science Publications: Oxford University Press.
Pashley D W (1956) Adv. Phys. **5** 173.
Schlenker, M, Buffiere, J-Y, Maire, E and Peix, G (1999) J Phys. D **32** A145.

Davis, T J, Gao, D, Gureyev, T E, Stevenson, A W and Wilkins, S W (1995) Nature **373** 595.

Stranski, J N and Krastanow, L (1938) Ber. Akad. Wiss. Wien **146** 797.

Whatmore, R W, Goddard, P A, Tanner, B K and Clark, G F (1982) Nature **299** 44.

CHAPTER 2

AN INTRODUCTION TO X-RAY SCATTERING

2.1. The interaction of X-ray photons with the sample

The X-ray photon interacts with the sample in many different ways and the form of interaction depends on the photon energy and the nature of the sample. X-ray photons are electromagnetic and it is the electric field vector that interacts most strongly with the sample. The magnetic interaction is small and is only observable under special conditions with very intense X-ray sources. There are several forms of interaction depending on the photon energy and the nature of the electron state. Electrons loosely bound to atoms, for example the valence electrons, may absorb part of the energy of a photon and the emitted photon will have a lower energy and longer wavelength. If it is assumed that the electron is stationary and totally unbound then this wavelength change is given by;

$$\Delta\lambda = \lambda_{SCATTERED} - \lambda_{INCIDENT} = \frac{h}{mc}(1 - \cos 2\omega') \qquad 2.1$$

This basically reflects the kinetic energy taken up by the electron. This interaction is termed Compton scattering. The wavelength change is therefore independent of the wavelength of the incident photon but varies with scattering angle, $2\omega'$, and is small (~0.024Å at most). An electron is not stationary or totally unbound in a solid and this will influence the energy (and wavelength) spread of the scattered photon. This makes the Compton scattering process a very useful tool for studying electron momenta in solids, etc. Because the wavelength change is so small, typical X-ray detectors used in diffraction experiments cannot discriminate this contribution from elastic scattering processes, therefore Compton scattering appears as a background signal. Each photon involved in this process will scatter independently. The scattering probability of coherent and Compton scattered photons for any given atom are of the same magnitude. However waves scattered in phase redistribute this intensity into sharp maxima that give intensities approximately related to N^2 (where N is the number of contributing atoms) compared to N for Compton scattering. Since N is generally very large the Compton scattering contribution is negligible, unless we are dealing with samples of very poor crystallinity. Equation 2.1 indicates that the scattering along the incident beam direction is zero. Compton scattering therefore increases with increasing scattering angle.

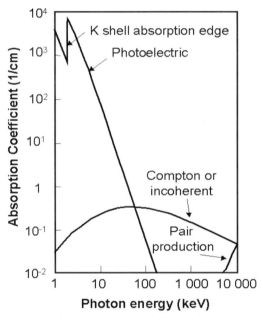

Figure 2.1 The various ways in which X-rays interact with matter as a function of their energy.

The tightly bound electrons will appear as a large immovable mass to a photon and therefore the energy transfer on interaction is very small (Rayleigh scattering), that is, the quantised energy state of the electrons is unchanged. If the wavelength of the photon is greater than that of the energy levels in the atoms of the sample then photoelectron absorption can occur and this is particularly strong when the energy exactly matches one of these energy transitions. For photon energies greater than a transition the electron takes up the remaining energy. This is the true absorption process when the photon is lost, although depending on the existence of free energy levels, the recovery of the electrons to lower energy levels can result in X-ray emission (fluorescence) or by involving an extra transition and electron emission (Auger process). Figure 2.1 shows the dominant absorption processes for different energy photons. Clearly at high energies the generation of electron-hole pairs increases and incoherent scattering rises to a maximum at 100keV in Si, whereas photoelectric absorption dominates at low energies. These low energy photons are primarily scattered by the localised electrons and each interaction will represent an instantaneous snapshot of the atomic positions. The temperature-dependent vibrational frequency of the atoms about their average site is many orders of magnitude lower than the sampling time of the X-ray photon: this has important consequences on how the intensity is averaged. An incoherent scattering process is one in which the emitted photon has a significantly different energy from the incident photon and the coherent process corresponds to the case where the energy of the emitted and the incident photons are the same. One very useful incoherent process is the generation of fluorescent radiation whose energy is element specific and therefore a very useful chemical analysis tool. However the coherent scatter or elastic scattering is the main emphasis of this chapter since this gives us access to the structural information of materials and their molecular configuration. At the low energies given in Figure 2.1, the wavelengths are comparable to the interatomic distances and provide a very useful probe of these lengths.

2.2 The nature of the scattered X-ray photon with no energy loss

We can consider the source of X-rays as a provider of photons that have a distribution of directions and energies. The location of each photon from the source to the sample and on to the detector is indeterminate and therefore a coherent relationship is maintained between all the possible paths to the detector. The possible paths are defined by the collimation between the source and sample and those between sample and the detector and from the interaction of the photon with the sample. These collimators could be crystals or slits depending on the application. However in this chapter we are concerned with the interaction of a photon with the sample under investigation. The instrument used to carry out the experiment will then average the contributions of all these photons and create the observed diffraction pattern. During the diffraction process the photon has the probability of several paths such that it exists along all probable paths until it is detected. What this means is that the probing X-ray photon is phase coherent over all its probable paths. The photon will now average over these probable paths and therefore the scattering will reflect the average of the regions in the sample that are probable paths.

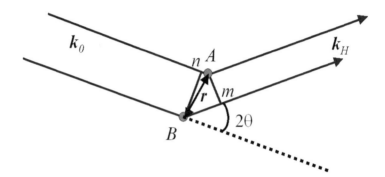

Figure 2.2 How two possible beam paths give rise to a phase difference characteristic of the separation of the scattering centres.

To understand coherence it is important to realise that a photon can only interact with itself, which cannot be understood in the classical sense but only with an understanding of quantum theory. If we therefore consider a photon to exist as a spatial distribution of probable paths then if all paths undergo different influences before recombining then interference can occur due to the fact that there is always a phase relationship between the same photon. If the paths cannot be recombined or cannot exist at the same time then there is no phase coherence. The distances over which phase coherence can exist is often termed the coherence length. *This differs from the correlation length, which is a sample dependent parameter and relates to the distances over which a phase relationship can be maintained.* Outside this region the photon paths cannot easily recombine, due to large orientation effects or phase averaging effects when the phases combined with random relationships; this effectively creates a null or noisy signal at the detector.

Consider the scattering of X-rays from two electrons separated by a distance r, figure 2.2. Suppose an incident wave with wave vector k_O of magnitude $1/\lambda$ impinges on two electrons A and B, where r is given by:

$$r = u\mathbf{a} + v\mathbf{b} + w\mathbf{c} \qquad 2.2$$

$\mathbf{a}$, $\mathbf{b}$ and $\mathbf{c}$ are unit cell translations and u, v and w are integers. If the electrons are set into vibration and excited they will emit secondary radiation defined by k_H of magnitude $1/\lambda$, then there will be a path difference between the scattering from A and that from B given by:

$$\text{An} - \text{Bm} = \lambda(\mathbf{r}.\mathbf{k}_0 - \mathbf{r}.\mathbf{k}_H) = \lambda \mathbf{r}.\mathbf{S} \qquad 2.3$$

$\mathbf{S} = (\mathbf{k}_O - \mathbf{k}_H)$ is the scattering vector. For the waves from A and B to be in phase then $\mathbf{r}.\mathbf{S}$ must be integer and hence:

$$(u\mathbf{a} + v\mathbf{b} + w\mathbf{c}).\mathbf{S} = integer \qquad 2.4$$

Since this must be true for all u, v and w then
$\mathbf{a}.\mathbf{S} = h$
$\mathbf{b}.\mathbf{S} = k \qquad 2.5$
$\mathbf{c}.\mathbf{S} = l$

These are Laue's equations, where h, k and l are integers. Suppose we take the first two equations, rearrange them and subtract them from each other

$$\left(\frac{\mathbf{a}}{h} - \frac{\mathbf{b}}{k}\right).\mathbf{S} = 0 \qquad 2.6$$

Therefore $(\mathbf{a}/h - \mathbf{b}/k)$ must be perpendicular to $\mathbf{S}$, and similarly for all the combinations. We can now consider a plane that intercepts the 'a' axis at $1/h$, the 'b' axis at $1/k$ and 'c' axis at $1/l$, which we can denote with Miller indices (hkl). Equation 2.6 can only be satisfied if $\mathbf{S}$ is orthogonal to the plane *(hkl)*, figure 2.3. The spacing between these sets of crystal planes containing the electrons is the projection of *a/h* on **S** (i.e. the distance between the plane and the origin), therefore

$$d_{hkl} = \frac{a.S}{h|S|} = \frac{1}{|S|} \qquad 2.7$$

by combining with Laue's equation. If the scattering angle 2θ, for the phase coherent scattering condition, is defined as the angle between k_O and k_H from figure 2.4

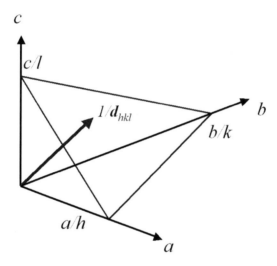

Figure 2.3 The plane (designated *hkl*) has a characteristic vector that has a length given by the inverse of its distance from the origin and is normal to this plane.

$$|S| = \frac{1}{d_{hkl}} = \frac{2\sin\theta}{\lambda} \qquad 2.8$$

or

$$2\theta = 2\sin^{-1}\left\{\frac{\lambda}{2d_{hkl}}\right\} \qquad 2.9$$

This is Bragg's equation. From this approach we can therefore consider the scattering process as a reflection.

Suppose that we have a crystal, then a vector of length $1/d_{hkl}$ can represent each plane *hkl* with a direction along the plane normal. The end points of all these vectors will form a periodic array that relates to the periodicity of the crystal. This is a very convenient representation of the crystal and from equation 2.9 we can see that this lattice in "reciprocal space" will only be a small distortion of its image in "diffraction space" over small regions. This concept will be developed through the course of this chapter and discussed in section 3.2.

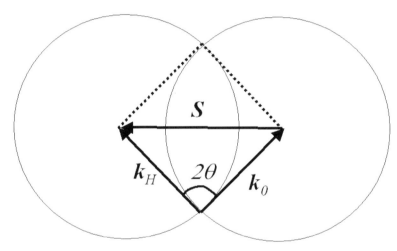

Figure 2.4 The relationship between the incident and scattered wave-vectors with the scattering vector.

So far we have considered the scattering from individual electrons, whereas in reality the electrons are distributed about the atomic positions. Also the atoms will vibrate about their average position and these factors must be included. Consider first that the electron cloud associated with the atom occupies a simple sphere then scattering from two different regions will interfere, figure 2.5. Debye (1915) showed how this can give rise to diffraction effects in gases and Compton (1917) indicated how atomic sizes could be estimated based on these ideas. The phase difference will therefore reduce the resultant scattering as the scattering angle $2\omega'$ increases. The strength of the scattering from an atom is therefore proportional to the number of electrons, which is further modified by this phase effect that depends on the size of the electron cloud. The ratio of the strength of the scattering from an atom to that of an individual electron is the scattering factor, which is given by the expression:

$$f(S) = \int \psi_2^*(r)\exp(iS.r)\psi_1(r)dV \qquad 2.10$$

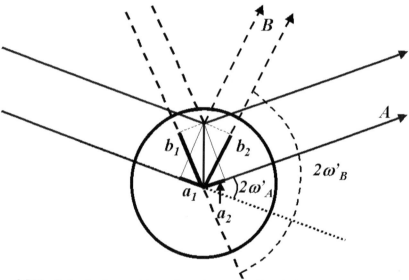

Figure 2.5 The finite atomic scattering volume (represented by circle) gives rise to a phase difference of possible beam paths that varies with scattering angle (compare beam paths A and B).

This equation does not include relativistic effects and is therefore not valid for very high energy (high frequency) X-rays and becomes less reliable in accounting for the scattering from the inner electrons of heavy atoms. ψ_1 and ψ_2 represent the initial and final states of the wave-function, but since we are concerned with coherent scattering these are the same. We can therefore consider the scattering factor (form factor) to be the Fourier transform of the charge distribution ($\psi^*\psi$) around an atom. The effective charge distribution is modified in the relativistic case, (Hartree, 1935; Fock, 1930). Most calculations of the scattering factor are based on a spherical approximation by averaging the various orbital contributions to effectively create that for a free atom or ion. Many authors have now calculated these scattering factors and to an excellent approximation they can be represented by an expansion of exponential terms

$$f_i = \sum_{j=1}^{4} a_j exp\left(-\left\{\frac{b_j sin\theta}{\lambda}\right\}^2\right) + c_i \qquad 2.11$$

The scattering factor for each atom in various states of ionisation can therefore be determined from a tabulated set of values (International Tables for X-Ray Crystallography **IV** p71 (1968)). The scattering factors for asymmetric distributions have also been calculated but for ease of calculation we will assume the spherical distribution is a good approximation.

So far we have considered the atoms to be static, whereas in reality the finite temperature of any sample will mean that they vibrate, section 2.7.4. Since X-rays are very high frequency electromagnetic waves the sample will appear static to each photon, although the atoms in general will be displaced from their average positions. These displacements will create a phase difference between scattering atoms and can therefore be treated in a similar manner to the scattering factor. We are only interested in the vibrations parallel to the scattering vector, however if we assume that the vibrations are essentially isotropic then they will influence all intensity maxima in a similar way. We can then define a single "temperature factor" influence to the scattering:

$$f = f_i exp\left(-\left\{\frac{8\pi^2 <u^2> sin^2\theta}{\lambda^2}\right\}\right) \qquad 2.12$$

Where $<u^2>$ is the root mean square displacement of the atoms, based on the geometrical model of Debye (1914). Again some of these values ($8\pi<u^2>$) have been calculated or determined experimentally for various structures since these values are sensitive to the molecular environment,

Reid (1983). Many semiconductor materials have small amplitudes of vibration because of the strong sp³ hybridised bonding within an extended lattice. For typical semiconductors a common isotropic value applicable to all atoms in the structure is generally sufficient for modelling the intensities. However for flexible molecules in a lattice dependent on van der Waals or hydrogen bonding then this temperature factor can be highly anisotropic and very different for each atom site.

Let us now consider the validity of the geometrical description of diffraction. Bragg's equation is only approximate because the measured scattering angle is outside the crystal, whereas the actual scattering angle is between incident and scattered waves inside the crystal that are modified by the refractive index. The refractive index will vary with scattering angle and the angle between the entrance surface and the exit surface of the photon. This can be understood most simply on the understanding that the wavelength is modified in the sample and the Bragg angle must change to compensate, since all other parameters in the Bragg equation are invariant. We will see later that the refractive index is less than unity, i.e. the wavelength of the photon is larger than in vacuo.

From Bragg's equation we can see that the minimum interplanar spacing that we can measure is determined by the wavelength, i.e. $\boldsymbol{d_{hkl}} > \lambda/2$ hence for a typical X-ray wavelength of 0.15nm the smallest interplanar spacing measurable is 0.075nm, which is less than all interatomic bond lengths. Shorter wavelength X-ray sources closer to half these values are commonly used for molecular structure determination to obtain atomic scale resolution with ease. Measurements of thicknesses in multi-layer composites structures are also possible with these wavelengths but require very high angular resolution and in the limit a good understanding of the shape of the scattering profile. The limit to the maximum dimension that can be determined depends on the width of the diffraction profile which is composed of the smearing effects of the diffractometer (the instrument probe dimensions) and the intrinsic scattering width which is a function of the sample and scattering conditions. As we will see a typical intrinsic scattering width for a perfectly crystalline sample is of the order of seconds of arc (~0.001^0). This permits the measurement of lengths up to many microns.

At this stage we can see that the interplanar spacing within the sample can be determined from the positions of the scattering maxima through Bragg's equation, however the strength of these maxima depends on the internal structure. Clearly scattering planes that have high electron density will scatter more strongly than those with low electron density. Also when the scattering is stronger the probability of a photon penetrating any great depth is diminished. As we shall see, the width of the scattering maxima depends on the depth of penetration in perfect materials. We shall first consider the scattering from an ideally perfect crystal to illustrate the scattering process.

The scattering vector given in equation 2.3 and 2.8 and figure 2.4

$$S = k_h - k_0 \qquad 2.13$$

can be represented graphically as two spheres of radius $1/\lambda$ with their centres separated by S, since the magnitude of each wave vector is $1/\lambda$, figure 2.4. The magnitude of the scattering vector is $|S| = 1/d_{hkl}$. The points of intersection represent conditions when Bragg's equation is satisfied. Clearly if d_{hkl} is too small the spheres do not overlap. If they do overlap then any intersection results in a singularity that suggests that the physics at this point will require a more exacting description. The second point to recognise is that the average refractive index is less than unity, the wave-vectors within the crystal therefore have different magnitudes since their wavelength is longer. We therefore have two concentric spheres associated with each end of the vector S representing the internal and external wave-vectors.

We will consider three conditions:

1. The dynamical diffraction model that best describes the physics of the waves within the sample.
2. The kinematical diffraction model that considers the diffraction broadening to be large compared with the intrinsic diffraction width, applicable to weak scattering from small crystals.
3. Specular scattering where no wave is excited in the sample, but the X-rays are reflected from changes in electron density.

2.3. The near exact theoretical description of scattering

Consider a sample with a series of atomic planes that are parallel to each other, figure 2.6. If we considered the beam path AB of a photon to be at an incident angle, θ_{hkl}, to scatter it in the direction BC then it will also be at the correct incident angle to be scattered from the underside of these atomic planes. Immediately we can understand that we have a complex wave-field building up in the sample where energy is swapping back and forth between the incident and scattered directions. We therefore have a dynamic situation where the energy of the incident beam is diminished with depth due to losses to the scattered beam and interference between the multiple scattered beams interfering with each other along the direction parallel to AB. This is the basis of the model proposed by Darwin (1914a and b) and can be understood on purely geometrical terms. The scattering model that accounts for this is the dynamical scattering theory. Additional energy losses from the X-ray wave-fields arise from photoelectric absorption, Prins (1930).

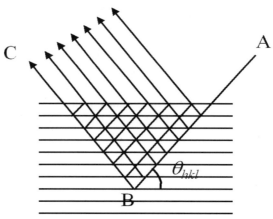

Figure 2.6 The complex interaction of X-rays with a perfect parallel-sided set of scattering planes.

To understand how the photon is scattered and generates an internal wave-field we can follow the physical description of Ewald or Laue. Ewald considered each atomic site to be occupied by a dipole that is set into oscillation by a passing photon. Each oscillating dipole emits radiation that adds to the total radiation field. We therefore have an array of dipoles oscillating which Ewald called "dipole-waves" all emitting electromagnetic radiation that interacts with other dipoles. The whole problem is considered by relating the resultant "dipole-wave" field to the electromagnetic field giving rise to it. Each dipole on a plane is assumed to emit in phase producing two plane wave-fronts in opposite directions normal to the atomic plane. We therefore have two wave types: an electromagnetic wave that is created by the dipole wave and the dipole wave itself. The "dynamical" aspect of this physical model is that the wave-field created by the dipole oscillations should be just sufficient to maintain them.

However we can see from Ewald's model that the dipoles are located at the atom sites, where in reality the sample has a distributed electron density and should be considered as a dielectric. Laue took this latter approach and the results are essentially the same as those determined by Ewald. We consider the crystal to exist of a continuous negative charge with shielded positive charges (the atomic nuclei) in a periodic array. When no incident photon exists, any position in the crystal can be considered as neutrally charged. But when an electric-field is applied there will be a relative displacement of the charges resulting in an electric polarisation and therefore the induced electric field (electric displacement) D is the resultant of the applied electric field E and the polarising field P

$$D = E + 4\pi P \qquad 2.14$$

If the electric field strength is not too strong then the induced electric field will be proportional to the local electric field strength. A strong electrostatic field can be defined as a significant proportion of the ionisation potential of an atom on the atomic scale. Since we are considering high energy electric fields associated with X-ray frequencies we cannot consider a simple dielectric constant of proportionality, where the electron density is isotropic. We hence introduce a variable electron density $\rho(r)$. If we assume that the restoring force on the electron is given by

$$m\frac{d^2x}{dt^2} = -eE \qquad 2.15$$

The electron displacement from its average undisturbed site is

$$x = -\frac{e}{m(\omega_0^2 - \omega^2)}E \qquad 2.16$$

where $\omega = 2\pi\nu$ and ν is the oscillation frequency of the X-rays and ω_o represents the natural frequency of vibration for the electron. This equation represents the resonance condition when strong absorption of the X-rays can take place. Clearly an infinite displacement is unrealistic so we can add in an imaginary term ($i\delta\omega$) to the bracketed term of the denominator, this can be derived from assuming the damping is proportional to the displacement. This imaginary term can be included by assuming that the damping term can be related to the individual scattering factors from each atom type and charge state. This is clearly wavelength dependent. The consideration given above is for a single electron atom and this has a strong bearing on the values associated with ω_0. For a many electron atom the natural or resonant frequency can be derived from the energy of excitation of a K shell electron, for example, to an unoccupied outer shell. These outer shells of any reasonable size atom are very close in energy and therefore form a series of unoccupied states giving a band of energies. Consequently ω_0 is not single valued, but rather complicated. However equation 2.16 does illustrate the general form of the resonance effect. It is also worth noting that atoms within a sample will create an almost continuous band of possible frequencies, however the dominant scattering is from the innermost localised electrons that are less sensitive to the immediate atomic environment.

Let us now rewrite the scattering factor as

$$f = f_0 + f' + if'' \qquad 2.17$$

f_0 is the term expressed in equation 2.11. Since we are mainly considering modifications to the scattering from inner shell resonance conditions the normal reduction of scattering with angle is very small for these two additional terms. The term f' is a direct modification to the scattering strength due to the proximity of the resonance condition and f'' introduces a phase lag in response to the incident X-rays. As the incident X-ray frequency moves further from this condition of resonance the electrons appear less tightly bound and behave rather like free electrons. Therefore the resonant frequency appears weak.

The polarising field will therefore relate to the product of the displacement, the charge and density of the charge carriers, i.e. electrons

$$P = xe\rho(r) = -\frac{e^2}{4\pi^2 mv^2}\rho_e(r)E \qquad 2.18$$

Therefore the induced electric field is given by

$$D = E(1+\chi) \qquad 2.19$$

where

$$\chi = -\frac{e^2\lambda^2}{\pi mc^2}\rho_e(r) \qquad 2.20$$

from equations 2.14, 2.18 and 2.19 and $v = c/\lambda$. This equation is equally applicable to the atomic nucleus but with its high mass, m, we can see that its influence on the electric field is close to a factor of 2000 times weaker than that from electrons. χ is the effective polarisability or the electric susceptibility of the crystal and will therefore vary throughout the crystal with the same periodicity as the electron density

$$\rho_e(x,y,z) = \frac{1}{V}\sum_{h=-\infty}^{\infty}\sum_{k=-\infty}^{\infty}\sum_{l=-\infty}^{\infty}F_{hkl}\exp(-2\pi i(hx+ky+lz)) \qquad 2.21$$

where x,y,z are the fractional position co-ordinates within the unit cell of volume V. The crystal can therefore be considered as a structure with an anisotropic periodic complex susceptibility. The assumption in using a summation and not an integral here is that the electron density is strongly associated with the atomic sites, i.e. the inner electrons dominate. Clearly from equation 2.21 the molecular structure in terms of the electron density distribution can be determined. However the summation is over all hkl i.e. all crystallographic planes. In practice the scattered intensity from only a finite number of "reflections" can be measured and this limits the resolution in determining the electron density. Since this assumption now strongly associates the scattering centres close to the atomic sites the physical model of the sample is equivalent to that proposed by Ewald (1916a, 1916b, 1918).

The quantity F_{hkl} is the structure factor for the unit cell or periodic repeat unit of volume V. The structure factor is the sum of all the scattering contributions from all the electrons in the unit cell

$$F_S = \int_r f_r \exp\{-2\pi i \mathbf{S}.\mathbf{r}\}d\mathbf{r} \sim \sum_{j=1}^{N} f_j \exp\{-2\pi i \mathbf{S}.\mathbf{r}_j\} \qquad 2.22$$

N is the number of atoms in the unit cell. The structure factor $\mathbf{F}_S$ in this equation is given in terms of the variable $\mathbf{S}$ and this is more general. The integral represents the full description assuming a distributed electron density. If however we assume the electrons are very localised the summation is adequate. The atomic size effect is accommodated by modification to the scattering factor as well as thermal vibrations. If we now define the location of atom j in terms of the fractional co-ordinates x, y and z within a unit cell of vectors $\mathbf{a}$, $\mathbf{b}$ and $\mathbf{c}$, equation 2.2

$\mathbf{r} = x\mathbf{a} + y\mathbf{b} + z\mathbf{c}$

then from Laue's equations

$\mathbf{r}.\mathbf{S} = x\mathbf{a}.\mathbf{S} + y\mathbf{b}.\mathbf{S} + z\mathbf{c}.\mathbf{S} = hx + ky + lz \qquad 2.23$

$\mathbf{F}_{hkl}$ will therefore be complex because it contains phase information associated with the atom positions within the unit cell. In describing the condition of resonance that dramatically changes the polarising field we established that a complex term could be included associated with the electron density. Since this resonance damping term is element specific the most appropriate way of including this is through the scattering factor, equation 2.17.

The electric field built up inside the crystal must obey Maxwell's equations

$$\operatorname{curl} \boldsymbol{H} = \frac{1}{c}\left(\frac{\partial \boldsymbol{D}}{\partial t} + 4\pi \boldsymbol{J}\right) \sim \frac{1}{c}\left(\frac{\partial \boldsymbol{D}}{\partial t}\right)$$

$$\frac{1}{c}\frac{\partial \boldsymbol{B}}{\partial t} = -\operatorname{curl} \boldsymbol{E} \sim -\operatorname{curl}\left\{\frac{\boldsymbol{D}}{1+\chi}\right\} \qquad 2.24$$

$$\operatorname{div} \boldsymbol{D} = 4\pi\sigma = 0$$

$$\operatorname{div} \boldsymbol{B} = 0$$

Where $\boldsymbol{H}$ is the magnetic field and c the velocity of the wave. The conductivity, σ, is assumed to be zero at X-ray frequencies therefore the current density J and the charge density are assumed to be zero. This also suggests that there will be no resistive heat loss. The approximations given here are those given by Laue (1931), where it is assumed that vectors $\boldsymbol{E}$ and $\boldsymbol{D}$ have the same direction. The displacement field, $\boldsymbol{D}$, the magnetic field, $\boldsymbol{H}$, and the wave vector $\boldsymbol{K}_m$ form an orthogonal set, whereas to be precise $\boldsymbol{E}$ lies within the plane of $\boldsymbol{D}$ and $\boldsymbol{K}$ and therefore can have a component along the propagation direction. Hence $\boldsymbol{D}$ forms a transverse wave and $\boldsymbol{E}$ does not. This introduces a small deviation at the 10^{-5} level, this is discussed later.

The concept of the theory is that the resulting field is the sum of plane waves, which can be defined as:

$$\boldsymbol{D} = \exp(2\pi i \nu t) \sum_m \boldsymbol{D}_m \exp\{-2\pi i[\boldsymbol{K}_m \cdot \boldsymbol{r}]\}$$
$$\boldsymbol{H} = \exp(2\pi i \nu t) \sum_m \boldsymbol{H}_m \exp\{-2\pi i[\boldsymbol{K}_m \cdot \boldsymbol{r}]\} \qquad 2.25$$

These represent the total electric displacement and magnetic fields at time t and position $\boldsymbol{r}$ for a total of m waves propagating in the crystal. The frequency of the electromagnetic wave is ν and the scattered wave vector, $\boldsymbol{K}_m$ satisfies Bragg's equation within the crystal, i.e.

$$\boldsymbol{K}_m = \boldsymbol{K}_0 + \boldsymbol{S}_m \qquad 2.26$$

Where $\boldsymbol{S}_m$ is a scattering vector inside the crystal and $\boldsymbol{K}_0$ is the forward-refracted beam. If we combine equations 2.24 and 2.19 we obtain

$$\text{curlcurl } \boldsymbol{D} = \text{curlcurl } \boldsymbol{E} + \text{curlcurl } \chi \boldsymbol{E} = -\frac{1}{c^2}\frac{\partial^2 \boldsymbol{D}}{\partial t^2} + \text{curlcurl}\frac{\chi \boldsymbol{D}}{1+\chi}$$

and from the standard vector relationship we can state that curlcurl$\boldsymbol{D}$ = graddiv$\boldsymbol{D}$ - $\nabla^2 \boldsymbol{D}$, and since div$\boldsymbol{D}$ = 0 we can now write

$$\nabla^2 \boldsymbol{D} - \frac{1}{c^2}\frac{\partial^2 \boldsymbol{D}}{\partial t^2} = -\text{curlcurl}\left(\frac{\chi \boldsymbol{D}}{1+\chi}\right) \approx -\text{curlcurl}(\chi \boldsymbol{D}) \qquad 2.27$$

The approximation is justified since $\chi \sim 10^{-5}$ for most materials. Now we can substitute $\boldsymbol{D}$ (equation 2.25) into equation 2.27, but we have to consider the expansion of $\chi \boldsymbol{D}$ first i.e.

$$\chi \boldsymbol{D} = \exp(2\pi i v t)\sum_n \sum_m \chi_{n-m} \boldsymbol{D}_m \exp(-2\pi i \boldsymbol{K}_m.\boldsymbol{r})$$

where $v = c/\lambda = ck$ and $k^2 = \boldsymbol{k}^2$. This substitution gives the following equation:

$$(k^2 - K_m^2)\boldsymbol{D}_m - \sum_n \chi_{n-m}[\boldsymbol{K}_m \times (\boldsymbol{K}_m \times \boldsymbol{D}_m)] = 0$$

The expression in square brackets can be rewritten using the vector rule:

$$[\boldsymbol{K}_m \times [\boldsymbol{K}_m \times \boldsymbol{D}_n]] = \boldsymbol{K}_m(\boldsymbol{K}_m.\boldsymbol{D}_n) - \boldsymbol{D}_n K_m^2$$

For a transverse wave $\boldsymbol{D}$ is orthogonal to $\boldsymbol{K}$ and therefore $\boldsymbol{K}_m.\boldsymbol{D}_n = 0$ and therefore this equation can be simplified to:

$$\frac{K_m^2 - k^2}{K_m^2}\boldsymbol{D}_m = \sum_m \chi_{m-n}\boldsymbol{D}_{n(\perp K_m)} \approx \frac{K_m^2 - k^2}{k^2}\boldsymbol{D}_m \qquad 2.28$$

k is the magnitude of the incident wave vector ($= |\mathbf{k}| = 1/\lambda = v/c$) and is very similar in magnitude to the wave vector inside the crystal. Referring back to the assumption of Laue concerning the use of the displacement field we can derive a similar expression for the electric field:

$$\frac{K_m^2 - k^2}{K_m^2} E_m = \sum_m \chi_{m-n} E_n \sim \frac{K_m^2 - k^2}{k^2} E_m \qquad 2.29$$

This is a slightly simpler equation but the differences are in the fifth order due to the small longitudinal component in the electric field E_n in the scattering plane. However the boundary conditions can be simpler and are often preferred in multiple beam diffraction calculations.

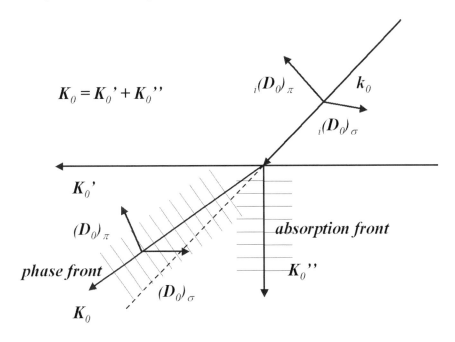

Figure 2.7 An incoming electromagnetic wave from a non-absorbing medium to an absorbing medium will create an absorption front traversing normal to the surface and a phase front along the propagation direction.

We can understand this latter point by considering a wave incident on a crystal close to the Bragg condition, figure 2.7. If we assume that the external incident wave is in vacuum and suffers no absorption then it is totally real and on entering the crystal it undergoes absorption. Since the component of the wave parallel to the vacuum / crystal boundary must be continuous the imaginary component of the wave vector must be perpendicular to the surface. The electric field component of the incident wave will be parallel to the displacement, since the susceptability is zero in an unpolarisable medium. The forward-refracted wave, because of its small deviation from the incident beam direction will now include a component of the electric field along the wave vector and is therefore partially longitudinal. However the transverse component is much more significant. The real part of the wave vector essentially defines the direction and the phase of the wave that varies along this direction. We can therefore consider that the regions of constant absorption are parallel to the surface and those of constant phase normal to the real component of the wave vector.

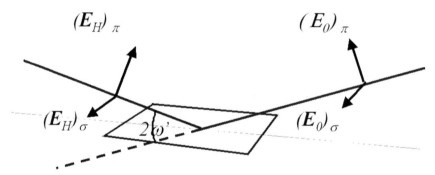

Figure 2.8 The change in the wave direction on scattering will influence the two electric field components to different extents depending on the scattering angle.

So far we have not considered the alignment of the transverse electric field direction and how this interacts with the sample. If the X-rays are generated by a laboratory source then the electric field will be circularly polarised, i.e. this direction is random but normal to the wave-vector. X-rays generated in a synchrotron have a very strong polarisation direction in the plane of the storage ring. We shall resolve the electric field polarisation into electric field vectors in the plane of scattering and normal to the direction, figure 2.8. The electric field component normal to the plane of scattering is unchanged apart from the reduction related to the reflecting power, the σ component. The electric field direction in the scattering plane, the π component is altered on scattering and effectively the electric displacement field is modified whenever it interacts with the scattering planes from above or below. The strength of the scattering from atomic planes is directly related to the susceptibility χ_{hkl} and χ_{-h-k-l}, and we can determine this reduction factor geometrically, figure 2.8

$$\{\chi_{hkl}\}_\pi = \chi_{hkl} \cos 2\omega'$$
$$\{\chi_{hkl}\}_\sigma = \chi_{hkl} \qquad\qquad 2.30$$

Clearly as the scattering angle moves close to the 90° the contribution of this π polarisation component becomes negligible compared with the σ component that is unchanged on scattering. Similarly as more reflections are involved in multiple crystal diffractometers the π component is reduced. We can now consider the equations above to be composed of two polarisation states and the scattered wave-fields are modified by a factor $C = \cos 2\omega'$ for the π component and $C = 1$ for the σ component.
We will now consider several conditions.

2.3.1. The condition of a single wave generated in a crystal

Suppose a wave is incident on a crystal and no diffracted wave is generated, then we have the displacement field at a position r to be given by

$$D = D_0 \exp(-2\pi i K_0 \cdot r) \exp(2\pi i \nu t) \qquad\qquad 2.31$$

Substituting this into our general equation 2.28 we obtain:

$$\frac{(K_0^2 - k^2)}{K_0^2} D_0 = \chi_0 D_0$$

i.e.

$$\{(1-\chi_0)K_0^2 - k^2\}D_0 = 0 \qquad 2.32$$

Since D_0 exists we have a simple expression:

$$|K_0| = \{1-\chi_0\}^{-1/2}|k| \sim \left(1+\frac{\chi_0}{2}\right)|k| \qquad 2.33$$

This solution just defines an average refractive index $(1+\chi_0/2)$ and the resultant dispersion surface is simply a sphere of radius $|K_0|$. This sphere is complex because the polarisability is complex, however a good representation and understanding can be obtained from considering just the real component. The excitation point on the dispersion surface is found by constructing a surface normal that intersects this sphere, i.e. at a point that creates a transmitted wave with a wave vector of length $|K_0|$, figure 2.9. The angle of the incident beam k_0 with respect to the surface normal is $\pi/2 - \omega$, we can therefore see that a second intersection point occurs on the larger diameter sphere. This second intersection point represents the specular reflected wave, with a wave vector k_R, and it is clear that this makes an angle of $\pi/2 - \omega$ with respect to the surface normal. This therefore represents a pure reflection with respect to the surface. The difference in the directions of the wave vectors K_0 and k_0 arise from the refractive index. This construction of the surface normal on the dispersion surface must and does satisfy the condition that the components of the internal and external wave vectors parallel to the surface are the same. This is a necessary and an obvious boundary condition.

From figure 2.9, we can see that the surface normal intersects each of these two concentric spheres in two positions. The incident wave creates a directly reflected wave and a transmitted wave. The other wave, dash line in figure 2.9 represents a wave that would be created by the transmitted wave being reflected; this becomes important for thin crystals when the propagating wave in the crystal reaches an interface. This wave has an increasing amplitude with depth. We can now see that we have two competing processes once the incident wave reaches the sample surface, that is the balance between the transmitted wave and the reflected wave.

If we consider the case when the surface normal only intersects the external sphere, line *CD*, then there is no internal wave vector and no wave field exists within the sample. This is the condition of total external reflection and for the geometric representation of the dispersion surface in figure 2.9 all the incident beam is specularly reflected. However the true dispersion surface is complex and it is this imaginary component that is associated with a finite absorption that results in a small exponential decaying wave-field to enter the sample.

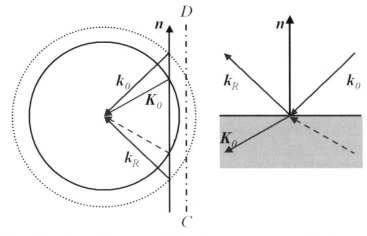

Figure 2.9 The dispersion surface construction for the one-wave case that just has a specular component.

It is clear from these arguments that specular reflection occurs throughout all ranges of angles but is only really significant at very low angles when the incident wave can be totally externally reflected. At this stage we can see that we are assuming some average refractive index. The method of calculating the intensity as a function of angle is considered very simply, section 2.8, using the more convenient optical theory. The specular profile emerges naturally from the dynamical theory by including all interactions with the dispersion surface.

2.3.2. The condition of two waves generated in a crystal

In this case we consider that the incident wave generates a scattered wave and therefore we have two waves of appreciable amplitude in the crystal:

$$D = \exp(2\pi i v t)\{D_0 \exp(-2\pi i K_0 . r) + D_m \exp(-2\pi i K_m . r)\} \quad 2.34$$

Hence substituting this into equation 2.28, for $m = 0$ and $m = H$, we have

$$\frac{K_0^2 - k^2}{K_0^2} D_0 = \chi_{-H} D_H + \chi_0 D_0$$

$$\frac{K_H^2 - k^2}{K_H^2} D_H = \chi_H D_0 + \chi_0 D_H$$

To include the influence of polarisation into these equations the susceptibility that relates to the strength of the scattering from a set of crystallographic planes, χ_H and χ_{-H}, are multiplied by the factor C. If these are rearranged and the effects of polarisation are included then

$$C\chi_{-H} D_H + \left(\chi_0 - \frac{K_0^2 - k^2}{K_0^2}\right) D_0 = 0 \quad 2.35$$

and

$$C\chi_H D_0 + \left(\chi_0 - \frac{K_H^2 - k^2}{K_H^2}\right) D_H = 0 \quad 2.36$$

For these two equations to have a common solution the determinant of these must be zero

$$\left(\chi_0 - \frac{K_0^2 - k^2}{K_0^2}\right)\left(\chi_0 - \frac{K_H^2 - k^2}{K_H^2}\right) = C^2 \chi_H \chi_{-H} \quad 2.37$$

This equation can be represented as two spheres whose surfaces are the loci of all the excitation points that can exist in the crystal at the chosen X-ray wavelength. This surface is the dispersion surface and is analogous to the Fermi or iso-energy surface of electron states in solids. This surface is given in figure 2.10 and is a shape of revolution about the scattering vector *OH*. The intersection of these spheres is no longer a singularity (simple cross-section) as in the simple geometrical understanding given above and this modification gives rise to a very good explanation of the profile shape and extinction effects that will be explained later.

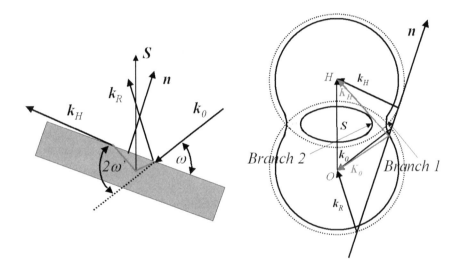

Figure 2.10 The dispersion surface for the two-wave case, indicating the removal of the singularity of the overlapping spheres. Both specular and scattered waves are generated.

The singularity is now a gap and represents the condition of total reflection; i.e. no wave-field exists in a non-absorbing crystal. The dispersion surface model greatly aids the understanding of the energy flow in a crystal. In general the flow is normal to the dispersion surface at the tie excitation points. As we increase the incident beam angle from a value below the Bragg condition the energy is gradually redirected from the forward-refracted direction towards the midpoint of the K_0 and K_H vectors. All the energy is then reflected and no waves are excited in the crystal. A further increase in the angle excites tie points on the lower branch (branch

2) of the dispersion surface and the energy flow moves back towards the forward-refracted direction. As the excitation points cross from one branch to another the phase of the scattered amplitude changes dramatically. Although the main concentration here is on the reflection condition when only one branch is excited at any time, for the transmission condition, when both branches are excited the phase difference can lead to very rapid changes in the amplitude depending on the scattering conditions (for example sample thickness). It is important to remember that the dispersion surface is complex and this takes account of absorption, i.e. perfect reflectivity is not possible and some sample penetration does exist.

Equation 2.37 is the satisfying condition for X-ray scattering and now we will derive the magnitude of the scattering as a function of incident angle. We take the abbreviation, see equation 2.28,

$$\beta_H = \frac{K_H^2 - k^2}{2K_H^2} \qquad 2.38a$$

that we call the deviation parameter.

The position of the surface normal as before can be used to describe the excitation points. The wave vector $K_H = k(1+\varepsilon_H)$, where ε_H is a very small value representing the relation between the external and internal wave vectors, figure 2.10. We can simplify equation 2.37, etc., with the approximation

$$2\beta_H = \frac{K_H^2 - k^2}{K_H^2} \sim \frac{K_H^2 - k^2}{k^2} \sim 2\varepsilon_H \qquad 2.38b$$

The assumption in this step is simply that $(\varepsilon_H)^2 \sim 0$ and as we shall see later $\varepsilon_H \sim 10^{-5}$ for X-ray wavelengths and typical materials under analysis, especially close to the scattering condition. This assumption makes the equation linear and more transparent. The complexity of the equation will be used later for the intermediate case where we will assume that the resulting approximation in the fundamental equation for dynamical theory (2.28) is sufficient, that is

$$\frac{K_H^2 - k^2}{k^2} = 2\varepsilon_H + \varepsilon_H^2 \qquad 2.39$$

This gives a perfectly workable solution, see later.

The boundary condition defines the surface normal with respect to the dispersion surface, as before. That is the component of the k_0 and the K_0 wave vectors parallel to the surface must be equal. We can therefore construct K_H as being composed of two components

$$K_H = k(1+\varepsilon_H) = k(1+\alpha_H) - kg\sin(\omega - 2\theta) \qquad 2.40$$

The parameter α_H is a geometrical factor that can be related to the angular deviation of the incident beam from the Bragg condition. The parameter g is the component of the vector $(k_0 - K_0)$ in terms of $|k|$. This is normal to the surface direction. Similarly the refracted wave vector can be written as

$$K_0 = k(1+\varepsilon_0) = k - kg\sin\omega \qquad 2.41$$

The parameter g is small and by using Snell's law for refraction and equation 2.33 we can derive an approximate magnitude.

$$g = \sin\omega - (\sin^2\omega + \chi_0)^{1/2} \approx \frac{\chi_0}{2\sin^2\omega} \qquad 2.42$$

Since the susceptability is small the second terms in both equation 2.40 and 2.41 are very small compared with the first terms. The geometrical factor α_H can be determined precisely using the construction in figure 2.11. The length $k(1+\alpha_H)$ is the distance from the tie-point intersection on the external wave vector sphere to the end of the scattering vector, i.e. PH. Using the cosine rule in triangle PHL_0 we have:

$$\alpha_H = \left\{1 + 4\sin\frac{\omega-\omega_0}{2}\left[\sin\frac{\omega-\omega_0}{2} + \sin\left(2\theta - \frac{\omega-\omega_0}{2}\right)\right]\right\}^{\frac{1}{2}} - 1$$

$$2.43$$

50 X-RAY SCATTERING FROM SEMICONDUCTORS

If we assume that the angular deviation is small then we can make the approximation

$$\alpha_H \sim 2\sin\frac{\omega-\omega_0}{2}\sin 2\theta \sim -(\omega-\omega_0)\sin 2\theta \qquad 2.44$$

ω_0 is the incident angle that satisfies the Bragg condition. The second approximation is that given in the original theory is inadequate for most purposes that we will consider.

We can eliminate the g parameter by introducing an amplitude ratio ($X = D_H/D_0$) combining equations 2.40 and 2.41 with 2.35 and 2.36. The resultant equation is given by:

$$\chi_{-H}\frac{\gamma_H}{\gamma_0}X^2 + \left\{\chi_0\frac{\gamma_H}{\gamma_0} - \chi_0 + 2\alpha_H\right\}X - \chi_H = 0 \qquad 2.45$$

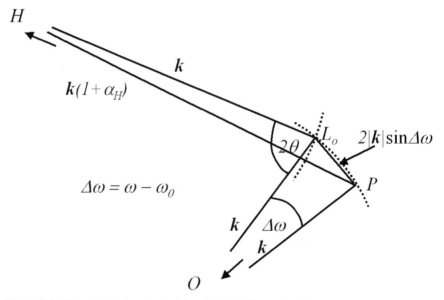

Figure 2.11 The construction for deriving the deviation parameter.

The sin($\omega-2\theta$) and sin(ω) have been substituted by γ_H and γ_0, these are the direction cosines of the incident and scattered wave vectors. A further assumption has been applied here in that the direction cosines of the internal and external wave vectors are the same. To be more precise we could use the direction cosines of the local internal wave vector with a refractive index correction. The corrected values of γ_H and γ_0 therefore become:

$$\gamma_0' = \left\{\frac{\gamma_0^2 + \chi_0}{1+\chi_0}\right\}^{\frac{1}{2}} \sim \left\{\gamma_0^2 + \chi_0\right\}^{\frac{1}{2}} \qquad 2.46$$

$$\gamma_H' = \left\{\frac{\gamma_H^2 + \chi_0}{1+\chi_0}\right\}^{\frac{1}{2}} \sim \left\{\gamma_H^2 + \chi_0\right\}^{\frac{1}{2}} \qquad 2.47$$

We can see that the direction cosines are strictly complex because of the polarisability. However the polarisability is of the order of $\sim 10^{-5}$ and therefore the overall influence to peak positions by the refractive index is an angular modification of 0.0066° at 5^0. The resulting correction in γ_0 is a mere 0.1%. At 3^0 incident angle this increases to 0.2%. These are very small effects since these parameters will only shift the peak positions by an additional and very small amount compared with the refractive index. Also as the angle of incidence or exit increases these corrections reduce.

The equation 2.45 can now be solved, since this is a standard quadratic and this gives:

$$\frac{D_H}{D_0} = X = \frac{-\eta \pm \sqrt{\left\{\eta^2 + 4\chi_H \chi_{-H} \frac{\gamma_H}{\gamma_0}\right\}}}{2\chi_{-H} \frac{\gamma_H}{\gamma_0}} \qquad 2.48$$

where

$$\eta = 2\varepsilon_H - \left\{1 - \frac{\gamma_H}{\gamma_0}\right\}\chi_0 \qquad 2.49$$

If we now include the non-linear aspect within the approximation given in equation 2.39 we obtain:

$$\frac{D_H}{D_0} = X = \left\{\frac{\mp\tau \pm \sqrt{\tau^2 - 4\left[\varsigma + \left(\frac{\gamma_H}{\gamma_0}\right)^2\right]\xi}}{2\left[\varsigma + \left(\frac{\gamma_H}{\gamma_0}\right)^2\right]\chi_{-H}^2}\right\}^2 - \frac{(1+\chi_0)}{\chi_{-H}} \qquad 2.50$$

Where:

$$\tau = 2(1+\varepsilon_H)\frac{\gamma_H}{\gamma_0} + \left(\frac{\gamma_H}{\gamma_0}\right)^2 \qquad 2.51$$

$$\varsigma = \frac{\chi_0 - (2\varepsilon_H + \varepsilon_H^2)}{\chi_{-H}} \qquad 2.52$$

$$\xi = \chi_H - \varsigma(1+\chi_0) - \tau \qquad 2.53$$

The equivalent deviation parameter is less clearly separated, but this derivation removes the main assumptions in the solution of the dynamical interaction of the case of two waves in a periodic medium. We therefore have a model that uses an exact geometrical factor, equation 2.43, by avoiding the straightforward linear approximation, equation 2.44, and takes into account the difference in direction cosines of the internal and external wave fields.

From these equations we can therefore calculate the scattering profile. The measured intensity at each position ω is now given by:

$$I(\omega) = X(\omega)X^*(\omega)\left|\frac{\gamma_H}{\gamma_0}\right| \qquad 2.54$$

The inclusion of the direction cosines accounts for the effects of beam compression or expansion. The original wave field energy relates to a unit area and therefore if there is beam expansion and all of the scattered beam is captured then we have to account for this.

2.3.3. A further discussion on the deviation parameter β_H

The full profile should be derived from all possible intersections with the dispersion surface, figure 2.10, which can be complicated further by multiple scattering events. However if we assume a single incident and scattered beam we can assume a deviation parameter. This assumption will only be valid far from the influence of strong specular reflectivity. The above derivation of the deviation parameter is clear to see using a purely geometrical approach however it may appear more transparent by visualising its derivation vectorially. It is important to remember that the surface normal that intersects the dispersion surface purely represents the components of the condition that the wave-vectors parallel to the surface outside and inside the crystal are equal. Hence we can rewrite the equation 2.38 as:

$$\beta_H \approx \frac{K_H^2 - k^2}{2k^2} = \frac{\{(_\parallel K_H + _\perp K_H) - (_\parallel k_H + _\perp k_H)\}\{(_\parallel K_H + _\perp K_H) + (_\parallel k_H + _\perp k_H)\}}{2k^2}$$

$$2.55$$

remembering of course that $k_H = k$, the external wave vector. Since the parallel components are equal this simplifies to:

$$\beta_H \approx \frac{_\perp K_H - _\perp k_H}{2k^2}(K_H + k) \sim \frac{_\perp K_H - _\perp k_H}{k} \qquad 2.56$$

Now consider figure 2.12, which represents the scattering vector and the incident and scattered wave vectors for the Bragg condition and a position deviated from this. From this figure we can see that

$$_\perp K_H = |S|\cos\varphi - |K_0|\sin\omega \qquad 2.57$$

$$_\perp k_H = -k\sin(2\omega'-\omega) \qquad 2.58$$

Hence

$$\beta_H \approx \frac{1}{|k|}\left[|S|\cos\varphi - |K_0|\sin\omega + |k|\sin(2\omega'-\omega)\right] \qquad 2.59$$

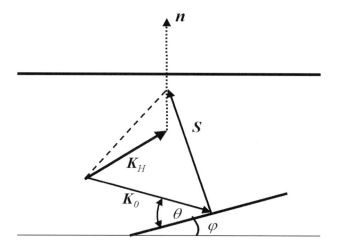

Figure 2.12 An alternative method of deriving the deviation parameter. This is incompatible with the dispersion surface construction.

Now the scattering vector is given by Bragg's equation 2.8, $|S|=1/d=\{2\sin\theta/\lambda\}(1+\chi_0)^{1/2}$, where the wavelength within the crystal has been modified with a refractive index term. The internal incident wave vector is refracted and modified similarly by this term. The magnitude of the external wave vector $k = 1/\lambda$. Substituting into equation 2.59, and taking the component normal to the surface we have

$$\beta_H \approx \{2\sin\theta\cos\varphi(1+\chi_0)^{1/2} - (1+\chi_0)^{1/2}\sin\omega + \sin(2\omega'-\omega)\}\gamma_H$$
2.60

If we neglect the refractive index contribution, since this is very small we have the expression derived by Zaus (1993). Although this is quite a different expression from those derived above it is important to realise here that the scattered beam direction must be known and it will be derived below in section 2.6. The major difference with this expression is clear to see in figure 2.12, compared with that of figure 2.10. Equation 2.43 relates more directly to the dispersion surface and the vectors are associated with the end points of the diffraction vector at the Bragg condition. This is a simple extension of the Laue and Ewald theories. The expression 2.60 allows the diffraction vector to move along the surface normal and is therefore less easily visualised in terms of the dispersion surface. However the diffraction profiles derived from these two approaches are indistinguishable provided that the γ_H and γ_0 parameters are associated with the Bragg condition for the individual distorted regions for the former case and are varied with actual angles in the latter. However there is disagreement when the strains are very large or the angular range modelled is large when γ_0 and γ_H vary significantly. When either of these conditions exist equation 2.60 should be used.

Equation 2.60 has been compared with simulations based on the solution of the full dispersion surface, Holy and Fewster (2003). The agreement is precise except close to grazing incidence when the specular component is significant or when the scattering from another reflection interferes coherently. These are special cases and for the analyses we shall discuss these are not a major concern.

So far we have been assuming that these derivations are based on an extended perfectly periodic crystal without distortions. There have been many attempts to extend the model to accommodate distortions and perhaps the most successful is that based on the concepts put forward by Takagi (1962). This will now be discussed.

2.4. A scattering theory to accommodate real crystals

The model above gives good agreement with experimental results but is rather restricted when applied to crystals that are distorted or consisting of multi-layers of different materials, etc. The conceptual approach first proposed by Takagi (1962) and expanded by Taupin (1964) and Takagi (1969) and others will be presented here.

From equations 2.20 and 2.21, the susceptibility at position *r* in the sample is given by:

$$\chi(r) = \sum_{S_m} \chi_{S_m} \exp(-2\pi i S_m \cdot r) \qquad 2.61$$

S_m being a reciprocal lattice vector and

$$\chi_{S_m} = -\frac{r_e \lambda^2}{\pi V} F_{S_m} \qquad 2.62$$

and

$$r_e = \frac{e^2}{mc^2} \qquad 2.63$$

i.e. the electron radius. A structure that is distorted can be considered to have an average and a locally perturbed susceptability, χ' due to a scattering centre displaced to the position co-ordinate *(r − u(r))*, hence

$$\chi'(r) = \sum_{S_m} \chi_{S_m} \exp\{-2\pi i (S_m \cdot r - S_m \cdot u(r))\} \qquad 2.64$$

$S_m \cdot r$ has the periodicity of the average lattice and is therefore rapidly varying on the scale of the interatomic spacings, whereas grad($S_m \cdot u(r)$) is assumed to be a variable on a larger macroscopic scale. Similarly the displacement fields can be given as:

Chapter 2 An Introduction to X-Ray Scattering 57

$$D(r) = \exp(2\pi i vt)\sum_m D'_m(r)\exp\{-2\pi i(K_m \cdot r - S_m \cdot u(r))\} \qquad 2.65$$

This sum of plane waves must satisfy Maxwell's equations as before and hence by deriving expressions for curlcurl$(D/(1+\chi))$, $\nabla^2 D$ and substituting into equation 2.27 we obtain:

$$\sum_m \left\{ (k^2 - K_m^2)D'_m - iK_m \cdot \mathrm{grad}(D'_m) + \sum_n \chi_{m-n}[D'_n]_m \right\} \exp(-2\pi i(K_m \cdot r - S_m \cdot u(r))) = 0$$
$$2.66$$

The term $[D'_n]_m$ represents the component of D_n perpendicular to K_m. To make the manipulation possible there are a few assumptions that are included to achieve this result; i.e. the electric displacement D and the polarisability P and their first derivatives are macroscopic in their variation. Hence the second order derivatives of D, first order derivatives of P (P being very much smaller than D) and the divergence of the gradient of the deformation strain field ($\nabla^2 S_m \cdot u$) are assumed to be negligible.

Equation 2.66 can now be integrated with respect to r over a unit cell repeat since $u(r)$ is assumed to be almost constant within these dimensions. This yields a more general result than that given in equation 2.28

$$s_m \cdot \mathrm{grad}(D'_m) = \pi i \frac{K_m^2 - k^2}{K_m^2} K_m D'_m - \pi i K_m \sum_n \chi_{m-n}[D'_n]_m \qquad 2.67$$

$(s_m \cdot \mathrm{grad}) = \partial/\partial s_m$ is simply a differential operator since s_m is a unit vector along the wave vector K_m. As with the former approach we can now solve this for the case of two significant waves in the sample, i.e. for $m = 0$ and H the forward refracted wave and the scattered wave

$$\frac{i\lambda}{\pi}\frac{\partial D'_H}{\partial s_H} = \chi_0 D'_H + C\chi_H D'_0 - \frac{K_H^2 - k^2}{K_H^2}D'_H \qquad 2.68$$

and

$$\frac{i\lambda}{\pi}\frac{\partial D_0'}{\partial s_0} = C\chi_{-H}D_H' + \chi_0 D_0' - \frac{K_H^2 - k^2}{K_H^2}D_0 \sim C\chi_{-H}D_H' + \chi_0 D_0' \qquad 2.69$$

where $|K_m| \sim 1/\lambda$ and the polarisation modification C has been included, equation 2.30. We can simplify these equations by introducing an amplitude ratio ($X = D_H'/D_0'$) similar to that in equation 2.48. Note the similarity of these equations to 2.35 and 2.36, The former were simply the limiting case of these equations 2.68 and 2.69. The parameters s_0 and s_H are the length vectors along the incident and scattered beam directions and can be substituted by t/γ_0 and t/γ_H respectively, where t is the depth into the crystal normal to the surface plane. We therefore obtain

$$\frac{dX}{dt} = \frac{\pi}{i\lambda}\left\{\frac{1}{\gamma_H}(\chi_0 X + C\chi_H - 2X\beta_H) - \frac{1}{\gamma_0}(\chi_0 X + C\chi_{-H}X^2)\right\} \qquad 2.70$$

This equation can be solved by integration after separation of the variables and use of partial fractions. The resulting equation gives:

$$X(Z,\omega) = \frac{SX(z,\omega) + i\{E + BX(z,\omega)\}\tan(GS[z - Z])}{S - i\{AX(z,\omega) + B\}\tan(GS[z - Z])} \qquad 2.71$$

where

$$A = C\chi_{-H}\frac{|\gamma_H|}{\gamma_0} \qquad 2.72$$

$$B = 0.5\left\{\chi_0\left(1 + \frac{|\gamma_H|}{\gamma_0}\right) - 2\beta_H\right\} \qquad 2.73$$

$$E = -C\chi_H \qquad 2.74$$

$$S = (B^2 - AE)^{1/2} \qquad 2.75$$

and

$$G = \frac{\pi}{\lambda \gamma_H}$$

The deviation parameter in this case is a little more complex than above since K' refers to a distorted wave vector. We should add an additional term into the deviation parameter β_H since $r = r_0 + u(r)$, the local incident wave vector will be changed to $K' = K - \text{grad}(S_H .u)$. The deviation parameter becomes

$$\beta_H' = \beta_H - \frac{1}{k}\frac{\partial}{\partial t}(S_H .u(r)) \qquad 2.76$$

This additional term can be neglected if we can consider our sample in terms of layers of isotropic material. The whole system is greatly simplified in this way by reducing the problem to layers sufficiently thin to be considered isotropic. In this way very complex multilayer structures can be modelled.

This theory leads to a fundamental difference with that of the plane wave dynamical theory case given in section 2.3. Equation 2.65 for the wave has an extra exponential term which allows the gradually variation in the displacement field with position. This essential difference predicts subtle details observed in topographic images that the earlier plane wave theories could not explain and were left to phenomenological approaches. This makes this theory the basis for most studies from statistical dynamical theory, to model defect scattering behaviour, through to simulating section topographs. Of course it will also work well in modelling perfect crystalline materials.

The amplitude ratio is now a function of a depth co-ordinate and the parameters refer to the local susceptability, geometrical factors and of course distortion. The co-ordinate z refers to some depth in the structure for which the amplitude ratio $X(z,\omega)$ is known and Z refers to a depth for which the amplitude ratio is required. The distortion is contained in the parameter

β_H that is given in equation 2.60 or by equation 2.43, where $\alpha_H \sim \varepsilon_H \sim \beta_H$. The resulting scattered intensity is therefore given by

$$I(\omega) = X(\omega)X^*(\omega)\left|\frac{\gamma_H}{\gamma_0}\right| \qquad 2.77$$

This derivation is a useful step because now we can analyse the scattering from multilayer structures very easily. However let us take the argument further by considering structures that have misfit dislocations (not an uncommon defect in semiconductor multilayers).

2.5. Scattering theory for structures with defects

From equation 2.71 it is simple to see that we can determine the scattering from layer structures. At some depth into the sample, for example the bottom of the substrate or wafer, the scattered wave-field will be insignificant and this gives a starting boundary condition for the recursive calculation. Determining the amplitude ratio at each interface will create the initial ratio for the layer above; when the surface is reached the amplitude ratio will be related to the intensity that can be measured by the detector.

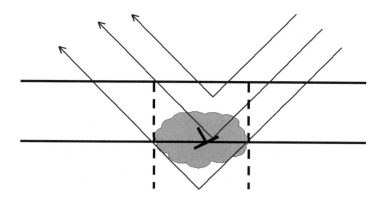

Figure 2.13 The different wave-fields that are created by the presence of a defect in the region of scattering.

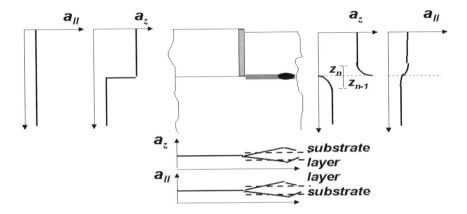

Figure 2.14 The variation of the distortion parallel and normal to the surface of a partially relaxed layer structure.

Consider now that at one of these interfaces the distortions are so great that the wave-field no longer sees an underlying periodic susceptibility but one that appears more like an amorphous region, Fewster (1992). We may then expect that the wave from the layer below will not couple into the layer above or at least not initially. We can then split the problem into the old wave field and a new wave-field, figure 2.13. An additional factor will be the distortion around a dislocation at an interface and this will add to the scattering. The form of this scattering can be very complex, but first of all let us try and visualise the strain state close to the dislocation core, figure 2.14. In the figure it is clear that if we assume a dislocation locally relieves the strain then the planes of atoms normal to the interface either side of this dislocation cannot go through a step change, this only happens at the dislocation. The alignment must occur over a finite distance either side of the interface. If we assume that some form of Poisson ratio is still applicable then the corresponding interplanar spacing parallel to the interface will take on a form given in figure 2.14.

This simple analysis of the strain will now give us a strain variation as a function of depth and can be modelled using equation 2.71. We can make an assumption that the variation of strain given in figure 2.14 follows an exponential form and that the lattice plane spacing parallel to the interface is some simple geometrical mean.

$$\bar{d}_{INTERFACE} = d_{n-1} + \frac{(d_n - d_{n-1})Z_{n-1}}{Z_n + Z_{n-1}} \qquad 2.78$$

The subscript n refers to the layer number before distortion and Z_n and Z_{n-1} refer to the distance of the distortion above and below the interface respectively. The distorted region is divided up into a series of thin layers of constant strain whose interplanar spacing is determined by an exponential distribution with end points $d_{INTERFACE}$ and d_n. The interplanar spacing perpendicular to the interface can therefore be determined from an assumed Poisson ratio for each thin layer.

Of course the structure of any distorted interface is unlikely to be uniform laterally and therefore we can split the problem into a series of columns, figure 2.15. This column approach can be applied to good approximation provided that the lateral dimensions are large compared with the depth, although a further modification, see section 2.7.3, can accommodate exceptions to this.

We should now consider the coherence associated with the possible wave-fields. As discussed at the beginning of this chapter we should consider that all possible paths of a photon are coherent if they can be emitted from the X-ray anode and arrive at the detector. Strictly therefore we should calculate all the path trajectories and bring them together coherently, however to reduce the calculation time we could bring in the idea of phase averaging. Phase averaging will lead to an effective incoherent addition provided that we model the major contributors to the scattering process. The intensity now becomes:

$$I(\omega) = [(X_a + X_b + \ldots)(X_a^* + X_b^* + \ldots) + X_c X_c^* + X_d X_d^* + \ldots]\left|\frac{\gamma_H}{\gamma_0}\right| \qquad 2.79$$

X_a and X_b represent the amplitude ratios from regions that scatter coherently with respect to each other and X_c and X_d represent regions that scatter independently and incoherently with respect to all other regions.

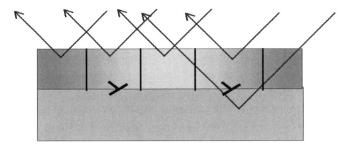

Figure 2.15 The procedure for generating the scattering from mosaic and distorted structures by splitting the problem into columns.

In this way we can now consider the scattering from layers below, within and above a distorted interface. In the opening argument we discussed that the susceptability no longer becomes meaningful and effectively the region close to the dislocation becomes amorphous. The strength of the scattering is effectively zero except at very small angles in specular reflection. We must therefore grade the susceptability as well as the strain. We could consider this as a very large static Debye-Waller factor, but that approach is only suitable for relatively small atomic misplacements (effectively within the harmonic limit of atomic vibrations). Here we are dealing with gross distortions.

Since we are concerned with columns that are relatively wide we have to average the structure within this width. If no correlation exist between these columns parallel to the surface then we can simply proportion the various contributions to the scattering, i.e.

$$I(\omega) = [(\Omega_a X_a + \Omega_b X_b + ...)^2 + (\Omega_c X_c)^2 + (\Omega_d X_d)^2 + ...]\left|\frac{\gamma_H}{\gamma_0}\right|$$

2.80

where

$$\Omega_a + \Omega_b + \Omega_c + \Omega_d + ... = 1 \qquad 2.81$$

Ω represents the proportion of the area occupied by a given amplitude ratio. Taking this approach it is clear that we can start to model quite complex

64 X-RAY SCATTERING FROM SEMICONDUCTORS

structures having both lateral changes in structure and in depth. We will take this a stage further although so far we have been taking a rather simple explanation of the scattering process as some pseudo one-dimensional effect. All we have been concerned with is the incident wave-field entering the sample being scattered somewhere and the detector producing a simple profile that we can model.

We shall now expand the method since extracting all this extra information requires a very different X-ray scattering experiment, that of collecting a reciprocal space map.

2.6. Scattering theory of reciprocal space maps

So far we have examined the scattering resulting from an incident wave creating a scattered wave but have not been concerned about the direction of the scattered wave. Suppose that we now only collect photons travelling along certain directions, i.e. our detector window is very narrow, then knowledge of the scattered wave direction is vital.

Consider figure 2.16 where we can once again use our knowledge of the boundary conditions. Since the parallel component of the scattering vector is constant across boundaries then it is constant with depth for any angle. Again we can write the scattering condition:

$$\boldsymbol{K}_H = \boldsymbol{K}_0 + \boldsymbol{S} \qquad 2.82$$

Since the parallel components are invariant

$$|\boldsymbol{K}_H|\cos(2\omega_i'-\omega_i) = |\boldsymbol{K}_0|\cos\omega_i + \frac{2\sin\theta_i}{\lambda}(1+\chi_0)^{1/2}\sin\varphi = |\boldsymbol{K}_0|\cos\omega_i + \frac{1}{d}\sin\varphi \qquad 2.83$$

The subscript i refers to the equivalent directions due to the refractive index effects. Similarly for the external parallel components that equate to those above we have:

$$|\boldsymbol{k}|\cos(2\omega'-\omega) = |\boldsymbol{k}|\cos\omega + \frac{2\sin\theta}{\lambda}\sin\varphi \qquad 2.84$$

The refractive index effects cancel, which is not surprising because now the external scattering angle is associated with a medium with a refractive index that cannot have imaginary components. Also the change in wavelength and Bragg angle within the crystal compared with the external values must compensate for each other since the interplanar spacing d is independent of the measurement method. The scattered wave direction is therefore

$$2\omega' = \cos^{-1}\{\cos\omega + 2\sin\theta\sin\varphi\} + \omega = \cos^{-1}\left\{\cos\omega + \frac{\lambda}{d}\sin\varphi\right\} + \omega$$

2.85

where $k = 1/\lambda$.

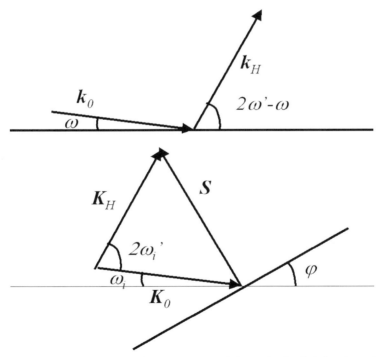

Figure 2.16 The internal and external wave-vectors and the angles for deriving the various relationships to satisfy the boundary conditions.

Hence now we have a more complete picture of the scattering process. As far as our experiment is concerned we have to ensure that the incident wave reaches the surface at the required angle and that the detector is placed at the correct angle to receive the scattered photon. The definition of the inclination of the scattering plane with respect to the surface is defined as in figure 2.16.

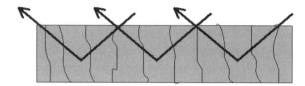

Figure 2.17 The procedure for generating the scattering from mosaic and distorted structures by splitting the problem into columns.

The scattering is now defined by two angles, ω and $2\omega'$, therefore the experimental data collection is two-dimensional. We will see in Chapter 3 on X-ray diffraction instruments how this data is collected but for the moment we will assume that we have an instrument that has an ideal detector acceptance and produces a perfectly well defined incident wave direction. Suppose now we have an ideal perfectly crystalline sample then the scattering will be confined to a line normal to the surface, this is realised through the equation 2.85 above.

For less than perfect structures where the crystallinity is not extended parallel to the sample surface as in figure 2.17 then the exit surface and for that matter the entrance surface to this crystal block may not be the sample surface. Effectively the incident beam passes through an absorbing medium, enters a finite crystalline block is scattered and the resultant wave exits through another surface and is absorbed before emerging at the sample surface. Since in general the crystal quality is less perfect, i.e composed of mosaic blocks that are not all identical, we can relax the rigor somewhat and consider averages and fluctuations. From figure 2.18 we can work out the scattering angle for a wave entering a crystal block, by applying the usual boundary conditions we obtain:

$$2\omega' = \sin^{-1}\{\sin(\omega - \zeta) - 2\sin\theta\cos(\varphi - \zeta)\} + \omega - \zeta \qquad 2.86$$

neglecting refractive index corrections. We have assumed for simplicity in this example that the block is parallel sided.

The broadening of the scattering profile due to the finite depth appears naturally in the calculations for the laterally infinite case, whereas the broadening due to the finite width is most easily estimated from the approximate description given in section 2.7.3. If we assume that the sides of the block are not parallel then the expression is essentially unchanged.

We can consider this by recognising that the incident wave enters the block and is subject to a small refractive index change, figure 2.10. The strength of the refractive index correction is dependent on the incident angle. The refracted wave, K_0, will then undergo scattering according to equation 2.82, and create an internal scattered wave with a wave vector of K_H. When this internal wave reaches the exit surface the boundary condition applies, which can be visualised by drawing the normal to the exit surface through the point of intersection of the refracted beam and the dispersion surface in figure 2.10. The intersection of this normal to the sphere centred on H of radius $|k_H|$ will denote the external scattered wave direction (this represents the condition when $\{K_H\}_{//} = \{k_H\}_{//}$). We therefore see that the exit surface defines the broadening direction of the scattered wave.

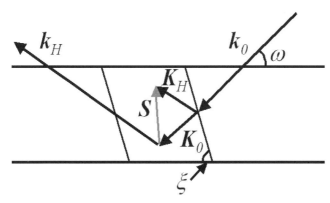

Figure 2.18 The components for deriving the relationships that satisfy the boundary condition for a simple case of a parallel-sided block.

To relate the coherence of the X-ray wave through different blocks we include changes in the phase due to the location. It is now possible to build

68 X-RAY SCATTERING FROM SEMICONDUCTORS

very complex structures from an array of blocks that represent some average of the local strain and scattering strength. We therefore have an additional lateral dimension and position to include in our model as well as the normal input of interplanar spacing, orientation and scattering strength, etc.

2.7. Approximate theory: The kinematical approach

Dynamical theory is based on a wave field approach: that is the incoming wave is attenuated due to scattering and also undergoes interference with the scattered wave. In the kinematical theory the incident wave is only assumed to be attenuated by normal photoelectric absorption. Therefore this theory is only strictly valid when the scattering is weak and is generally perfectly adequate to model the scattering from defects, ideally mosaic samples or highly distorted material. Another aspect that is ignored in the kinematical model is the refractive index. Of course care must be applied with the kinematical theory in that the total scattered intensity can exceed the incident wave intensity.

2.7.1. Comparison between dynamical and kinematical models of diffraction

We will firstly consider the differences between theories by removing some of the conditions associated with the analysis above. Consider for example the Takagi equations 2.68 and 2.69. The interference of the scattered wave with the forward refracted wave is a consequence of the magnitude of χ_{-H}, hence if we assume that the scattering strength is zero from the underside of the scattering planes then the real part of χ_{-H} is zero. This is sometimes referred to as primary extinction. The imaginary component is assumed to exist to account for normal photoelectric absorption. The refractive index arises from the finite magnitude of χ_0, therefore the real part of χ_0 is assumed to be zero. We can of course also remove the absorption effects, but in general this can easily be accommodated in the kinematical approximation from tabulated absorption coefficients. Therefore the equation 2.71 can be used in the kinematical approximation,

$$X(Z,\omega) = \frac{SX(z,\omega) + i\{E + BX(z,\omega)\}\tan(GS[z-Z])}{S - i\{AX(z,\omega) + B\}\tan(GS[z-Z])} \qquad 2.87$$

although A is purely imaginary and B becomes (the imaginary part of B) + β_H. Also the coupling of the wave fields between layers should be removed, just as in the case of a disrupted interface described in the section above. The thickness of the substrate will also create complications since the attenuation of the incident wave is greatly reduced and fringes created from the bottom surface strongly influence the pattern. The general approach is therefore to include the scattering from the substrate as dynamical and from the layers as kinematical. The influence of the refractive index in the dynamical model has to be removed for comparison. The differences between the dynamical and kinematical theories can be quite crucial and misleading in the analysis of multilayer structures. However for the analysis of weak scattering from small crystallites the kinematical theory can be perfectly adequate.

However if we include the assumptions of the kinematical approximation in equation 2.87 and take the assumption of an infinite crystal such that $\tan GS[z-Z] = i$, then

$$X(\omega(\sim \frac{-E}{2B} = -\frac{F_H}{(2\kappa 2_H - imag(F_0))} \qquad 2.88$$

where κ is some constant. We can see from this that it is only the absorption of the incident wave that prevents the amplitude becoming infinite at the Bragg angle (i.e. when $\beta_H = 0$). The intensity is then simply related to the square of the structure factor for the scattering position concerned. We can consider the kinematical model for an infinite lattice to create a scattering pattern that has delta functions at the Bragg angles with a magnitude related in some way to the square of the structure factor for the reflection concerned.

The information concerning the peak shape and true intensity is lost in the kinematical model. We can though include some form of shape function and impose this on the magnitude that we derive from the structure factors. As described above the dynamical and kinematical theories produce similar results for weak scattering. The intensities must therefore be less than ~10% of the useful incident intensity. However it is not possible to just measure the incident and scattered intensities and derive a

70 X-RAY SCATTERING FROM SEMICONDUCTORS

fraction and decide on the theory to use, except in very special circumstances. This will become clear in Chapter 3 on the instrumental techniques, when we consider the diffractometer divergence, etc.

2.7.2. The important derivations of the kinematical theory

Since we have established that this theory is based on weak scattering and crystals of limited extent in order to obtain realistic intensities, we shall lay out the appropriate assumptions. The intensity in a diffraction pattern from an infinite crystal is concentrated at the Bragg points and has a magnitude that is proportional to the square of the structure factors

$$I(S) = KF_S^* F_S \delta(s-r) \qquad 2.89$$

K is a constant of proportionality. Clearly this is significantly simpler than the dynamical model and can be a very powerful approximation for interpreting weak scattering.

Of course weak scattering is usually associated with limited small crystals and therefore we have to modify our model. Consider a sample with N scattering planes separated by distance d, figure 2.19. We can now write down the phase difference between scattering planes at scattering angles ε from the Bragg angle from deriving the path difference in the figure

$$\Phi = \frac{2\pi d}{\lambda}\{\sin(\theta+\varepsilon)+\sin(\theta+\varepsilon)\} \sim \frac{2\pi}{\lambda}\{\lambda+2d\varepsilon\cos\theta\} \equiv \frac{4\pi d}{\lambda}\{\varepsilon\cos\theta\} \qquad 2.90$$

for small deviations from the Bragg angle.

Chapter 2 An Introduction to X-Ray Scattering   71

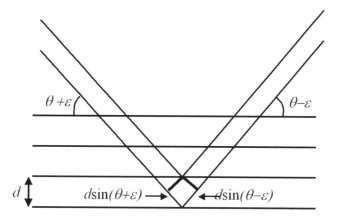

Figure 2.19 The influence on the path length and hence phase relationship for small deviations from the optimum scattering condition.

Suppose now we add all these contributions, then this will result in an amplitude A with a phase angle $N\Phi$. We can add these contributions using an Argand diagram. Each amplitude contribution, a, subtends an angle Φ to the centre of a circle of radius $a/\sin\Phi$ and the resultant is the vector sum of these having the same radius but phase angle $N\Phi$ and amplitude A. The radius for the resultant can be expressed as $A/\sin N\Phi$ where we can equate

$$\frac{A}{\sin N\Phi} = \frac{a}{\sin \Phi} \qquad 2.91$$

Therefore the resultant amplitude of the summation of all these wave-fronts is

$$A = a \frac{\sin N\Phi}{\sin \Phi}$$

The form of this equation is represented in figure 2.20. It is now possible to relate the depth over which scattering takes place and the thickness of the layer to the width of the scattered profile. The interatomic spacing gives the phase angle Φ from equation 2.90 and the layer thickness is represented by N interatomic spacings.

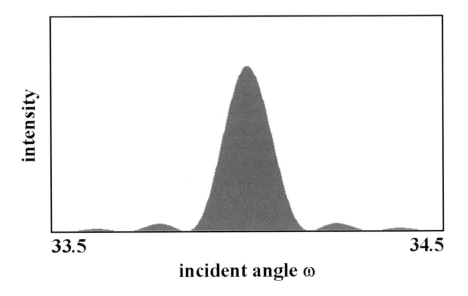

Figure 2.20 The shape of the interference function resulting from the finite size of the scattering region.

We can now make a further approximation by inspection of figure 2.20. The intensity of this profile clearly becomes small for $N\Phi$ larger than π therefore we can write:

$$(N\Phi)_{A\neq 0} = \frac{2\pi}{\lambda}\{2Nd\varepsilon_{A\neq 0}\cos\theta\} < \pi \qquad 2.92$$

The crystal dimension, L, or the region of coherent scattering is given by Nd, consequently we have

$$\varepsilon_{A\neq 0} \sim \frac{\lambda}{4L\cos\theta} \qquad 2.93$$

If the full width at half maximum intensity is measured on the scattering angle ($2\omega'$) scale then the depth of coherent scattering can be determined directly from this profile width from

$$L = \frac{\lambda}{\varepsilon_{FWHM\,(2\omega')} \cos\theta} \qquad 2.94$$

This is the Scherrer equation, Scherrer (1918).

So far we have ignored the aspects of absorption due to photoelectric effects and other non-scattering processes. We can consider the absorption to be related to the total path length of the X-rays in the sample. We will not consider the detailed processes here but rather take a pragmatic approach and consider the incident beam to loose energy (statistical loss of photon numbers) as it passes through the sample. If we have a small depth of material dt, with a density ρ that changes the incident beam intensity by $-dI$, we can assume that the loss of intensity is a proportional reduction and that the loss is proportional to dt. Hence we have:

$$dI = -\mu I dt \qquad 2.95$$

where μ is the total linear absorption coefficient. The ratio (μ/ρ), which we will term the mass absorption coefficient, is found to be roughly independent of the form of the material. Also to a good approximation the total losses are equal to the sum of the individual losses within the sample, i.e.

$$\frac{\mu}{\rho} = \sum_i \left(\frac{\mu}{\rho}\right)_i g_i \qquad 2.96$$

where g_i is the mass fraction of atom i. From the mass absorption coefficient and knowledge of the density of the sample we can derive the total linear absorption coefficient. Integrating equation 2.95 and defining the incident and transmitted intensity as I_0 and I, we have

$$I = I_0 \exp(-\mu t) \qquad 2.97$$

The mass absorption coefficients have a complicated form but have been determined and tabulated in detail in International Tables for Crystallography for each atom type for a large spread in wavelengths.

These mass absorption coefficients are proportional to the cube of the wavelength (Bragg-Pierce Law) except close to resonant absorption, section 2.3.

The calculation of the total attenuation in multi-layers must include these effects by including this attenuation through the whole structure above the scattering material. In periodic superlattices we can approximate this to

$$I = I_0 \prod_i \exp\left\{-\mu_i t_i \left(\frac{1}{\sin(2\omega' - \omega)} + \frac{1}{\sin \omega}\right)\right\} \qquad 2.98$$

Fewster (1986).

2.7.3. Lateral dimension analysis

In this section we shall extend the arguments given above to include lateral inhomogeneities into our scattering theory. Krivolglaz (1963) has developed an approach to include lateral correlation lengths by including what is effectively a Fourier transform of the length scales expected in the structure. This has been used to very good effect by many authors. However to make the derivation of lateral correlation lengths consistent with our arguments so far, we will consider a phase difference approach.

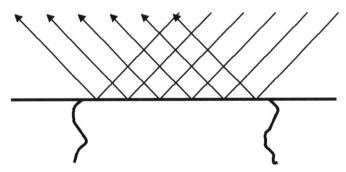

Figure 2.21 The scattering from a mosaic block is built by combining the amplitudes from an infinite number of possible paths.

Consider figure 2.21, where we have a mosaic block or some shape within our sample that can scatter along the whole dimension. If we consider the waves to be separated by some small dimension l then we can calculate the path difference between any two waves separated by l and derive the phase difference:

$$\Phi = \frac{2\pi l}{\lambda}\{\cos(2\omega'-\omega) - \cos\omega\} \qquad 2.99$$

The amplitude is again given by the function in equation 2.91. Of course ω relates to the incident wave-field direction and $2\omega'$ is given by equation 2.86. The dimension Nl is simply the dimension parallel to the surface plane of the sample. Now the separation of the individual wave-fronts contributing to this profile can be any small value and the shape is invariant. This is evident from the approximation that we can make for small phase angles:

$$a\frac{\sin N\Phi}{\sin\Phi} \sim a\frac{\sin N\Phi}{\Phi} = K\frac{\sin N\Phi}{\frac{2\pi}{\lambda}\{\cos(2\omega'-\omega)-\cos\omega\}} \qquad 2.100$$

where $K = a/l$ and is constant since the scattered contribution from a region of length l can be considered proportional to the length involved.

Therefore we can add the influence of mosaic blocks to our scattering pattern. The amplitude scattered by the layer, assuming that it is perfect should now be smeared out such that the amplitude is distributed to give measurable scattering at $2\omega'$ for a range of incident angles:

$$A = a\frac{\sin\{\frac{2\pi L}{\lambda}[\cos\{\sin^{-1}\{\sin(\omega-\zeta)-2\sin\theta\cos(\phi-\zeta)\}-\zeta\}-\cos\omega]\}}{\sin\{\frac{2\pi l}{\lambda}[\cos\{\sin^{-1}\{\sin(\omega-\zeta)-2\sin\theta\cos(\phi-\zeta)\}-\zeta\}-\cos\omega]\}}$$

$$2.101$$

for the case of a parallel sided block inclined at ζ to the sample surface. Clearly this approach can be extended to accommodate blocks of various shapes and different strain values.

76   X-RAY SCATTERING FROM SEMICONDUCTORS

This analysis is the situation that we would find for a series of identical mosaic blocks scattering with no phase relationship between them. We can address the situation of a distribution of block sizes by including contributions of several blocks as described in equation 2.80.

Figure 2.22 A laterally periodic structure will create a further modulation in the scattering pattern.

Extending these arguments further we can accommodate laterally periodic structures by considering the length scales laterally, l in equation 2.99 above, to become the wavelength of the periodicity. Figure 2.22 gives a simple grating with small hillocks periodically arranged over the surface. The smearing of the amplitude from a laterally finite structure is now composed of two terms

$$A = a \frac{\sin N_G \Phi_G}{\sin \Phi_G} \frac{\sin N_B \Phi_B}{\sin \Phi_B} \qquad 2.102$$

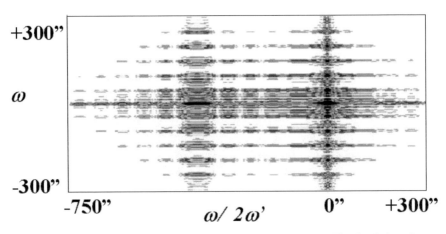

Figure 2.23 The complex pattern from a laterally periodic structure. The simulation of 0.3μm thick and 500nm wide AlAs hillocks repeated laterally 5 times with a period of 600nm on a GaAs substrate creates a complex pattern. This calculation is based on equation 2.102, dynamical theory and the discussion in section 2.6.

The periodic contribution from the grating G is modulated by the smoother interference function of the hillocks B. The dramatic effects are presented in figure 2.23. Of course here we are working with something which is perfectly periodic, yet we can include dispersion of the grating periodicity and hillock size by including several contributions to the amplitude term with different lengths:

$$\overline{A} = A_1 + A_2 + A_3 + A_4 + \ldots \ldots \qquad 2.103$$

It is obvious that we are now able to build up a very detailed model of our sample and with the combination of the dynamical and kinematical models to predict the scattering that we might expect.

2.7.4. Scattering by defects: Diffuse scattering

So far we have assumed the materials that we are studying are largely perfect, although we have included aspects of mosaicity and interfacial distortions. Of course materials are also likely to contain point defects (substitutional impurities, interstitial atoms and vacancies) and small dislocation loops, threading dislocations, etc. These features will influence the scattering pattern and will be considered in this section. The scattering is generally very weak and can be treated kinematically, although some authors have extended this to dynamical theory, (Kato, 1980; Olekhnovich and Olekhnovich, 1981; Holy, 1982; Khrupa, 1992; Pavlov and Punegov, 1998), but since we wish to give a general understanding the kinematical approximation will be discussed here. The differences in the two approaches are in the very fine detail that is usually obscured by the Bragg scattering, although of course it is very important to establish the validity of the kinematical theory approach and these articles largely confirm this.

Suppose we have a defect, this could be an interstitial or a vacancy for example, then the associated scattering will be related to the scattering factor for the defect multiplied by the number per unit volume. If the defects scatter independently (i.e. there is no phase relationship between them) then

$$F = \sum_{d=1}^{D} f_d \qquad 2.104$$

This will give rise to scattering that will be distributed throughout space with a total integrated intensity of

$$I = \left\{ \sum_{d=1}^{D} f_d \right\}^2 \qquad 2.105$$

For a low concentration of defects with intensity distributed throughout the scattering pattern this will be a rather insignificant background. Now the presence of a defect is also likely to displace the atoms of the matrix and this can be represented by a modification to the structure factor, such that equation 2.22 becomes

$$F_S = \sum_{j=1}^{N} f_j e^{-2\pi i S.(r_j + u_j)} \qquad 2.106$$

This represents a perturbation on the average lattice and in the kinematical approximation the intensity is given by

$$I = F_S^* F_S = \sum_n \sum_m f_n f_m \exp\{-2\pi i S.(r_n - r_m)\} \exp(-2\pi i S.(u_n - u_m)) \qquad 2.107$$

Now the atom movements can never be considered independent because of the bonding in a solid will result in a restoring force to any movement. This analysis is equally valid for thermal agitation or distributed defects although the distribution of stationary waves may differ and can be represented by a Fourier sum of phonon waves.

$$u_n = \sum_k A_k \cos\{\omega_k t - r_n.k - \delta_k\} \qquad 2.108$$

where **k** in this context is the wave-vector for a phonon wave with a maximum amplitude of vibration A_k along the direction of the scattering vector **S**. The phase term δ_k is to account for the differences observed by each photon during a measurement for a vibrating (thermally agitated) crystal lattice. Therefore we can now write

$$(u_n - u_m)^2 = \left[\sum_k A_k \cos\{\omega_k t - r_n.k - \delta_k\} - \sum_k A_k \cos\{\omega_k t - r_m.k - \delta_k\} \right]^2$$
2.109

Any static distortion due to defects can change the average separation in the atoms and this will appear as a change in the average lattice parameters and add to the Bragg scattering. We have less interest in this at the moment although it does offer a way of determining the defect density or alloy composition in semiconductors. Our purpose here is to examine the influence of the variations from the average. If we now determine the average of equation 2.109 we obtain

$$\overline{(u_n - u_m)^2} = \sum_k A_k^2 \{1 - \cos([r_n - r_m].k)\}$$
2.110

The time dependent and phase averaging are assumed to cancel on averaging and the average of $\sum_x \cos x = 0.5$. Now using the approximation for an exponential when the displacements are small,

$$\overline{\exp(-2\pi i S.(u_n - u_m))} \approx 1 - 2\pi^2 S^2 \overline{(u_n - u_m)^2}$$
2.111

Again we have removed those terms that just move the average, or average to zero, since these as described above just result in an expansion or contraction of the lattice structure.

80 X-RAY SCATTERING FROM SEMICONDUCTORS

$$\overline{\exp(-2\pi i \mathbf{S}.(\mathbf{u}_n - \mathbf{u}_m))} \approx 1 - 2\pi^2 S^2 \sum_k A_k^2 \{1 - \cos([\mathbf{r}_n - \mathbf{r}_m].\mathbf{k})\}$$

$$\approx \exp\left\{\frac{-16\pi^2 \sin^2\theta}{\lambda^2} \overline{u_S^2}\right\} \exp\left\{2\pi^2 S^2 \sum_k A_k^2 \cos([\mathbf{r}_n - \mathbf{r}_m].\mathbf{k})\right\}$$

2.112

We have substituted $S = 2\sin\theta / \lambda$, equation 2.8. If we compare this with equation 2.12 the first exponent in the left-hand side of equation 2.112 is the general temperature factor or Debye-Waller factor that reduces the intensity of the Bragg scattering. The factor of two arises from this equation relating to intensity whereas equation 2.12 relates to the scattering amplitude, i.e. $(\exp(-M)\exp(-M) = \exp(-2M))$. The displacement $\overline{u_S^2} = 0.5\left\langle \sum_k A_k^2 \right\rangle$, that is the average of half the maximum amplitude.

We now want the time average of equation 2.107, so substituting equation 2.112 into 2.107 we obtain

$$<I> = \sum_n \sum_m f_n f_m \exp\{-2\pi i \mathbf{S}.(\mathbf{r}_n - \mathbf{r}_m)\}\exp(-2M)\left\{1 + \left\langle 2\pi^2 S^2 \sum_k A_k^2 \cos([\mathbf{r}_n - \mathbf{r}_m].\mathbf{k})\right\rangle\right\}$$

2.113

The intensity is therefore composed of Bragg scattering with a reduction from a mixture of thermal effects and some deviation from perfection in the lattice and an oscillatory *cosine* term. Let us consider this *cosine* term in more detail.

These phonon waves create a modulation in the intensity rather similar to those of a synthetic modulated structure or superlattice described in section 4.4.4. This modulation will produce satellites centred on each Bragg peak at distances related to the phonon wavelengths that are relatively long. This becomes clear when we expand the *cosine* term

$$\exp\{-2\pi i S \Delta r_{n,m}\}\cos\left\{\frac{\Delta r_{n,m}}{\lambda_p}\right\} = \exp\{-2\pi i S \Delta r_{n,m}\}\frac{\exp\left\{\frac{i\Delta r_{n,m}}{\lambda_p}\right\}+\exp\left\{-\frac{i\Delta r_{n,m}}{\lambda_p}\right\}}{2}$$

$$= 0.5\left[\exp\left\{-2\pi i S \Delta r_{n,m}+\frac{i\Delta r_{n,m}}{\lambda_p}\right\}+\exp\left\{-2\pi i S \Delta r_{n,m}-\frac{i\Delta r_{n,m}}{\lambda_p}\right\}\right]$$

$$= 0.5\left[\exp\left\{-2\pi i \Delta r_{n,m}\left(S+\frac{1}{\lambda_p}\right)\right\}+\exp\left\{-2\pi i \Delta r_{n,m}\left(S-\frac{1}{\lambda_p}\right)\right\}\right]$$

2.114

This term is composed of a fixed term associated with the Bragg condition, S, and satellites appearing at a position inversely related to the phonon wavelength λ_p, figure 2.24.

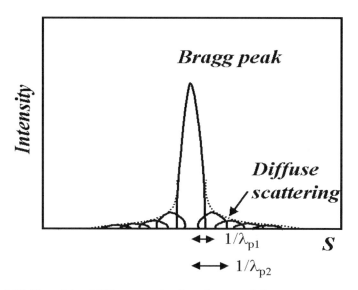

Figure 2.24 The origin of diffuse scattering from the sum of phonon waves.

The phonon wavelengths are large on the scale of the X-ray wavelength and will therefore produce broadening at the base of each Bragg peak. Also the distances over which these lengths are correlated and hence maintain an X-ray phase relationship will also add to the broadening of the scattering, equation 2.94. The correlation length for a short wavelength phonon would be expected to be small and therefore produce a weak broad profile, whereas the longer wavelength phonons have a longer correlation length and hence a sharper profile. The distribution of phonon wavelengths and their amplitudes will be a function of the defect or thermal vibration. In general therefore the intensity falls rapidly with decreasing phonon wavelengths and gives rise to the characteristic decline in diffuse scattering from the Bragg peak maximum characteristic of many experimental profiles, figure 4.14.

For weak deformations (e.g. small isolated elastic distortions) the intensity falls as $1/(\Delta S)^2$ and is characteristic of the scattering a long way from the core of a defect, Huang (1947). Closer to the defect core or for a low concentration of cluster defects the deformations will be strong and the intensity falls as $1/(\Delta S)^4$, etc. These fall-off rates are the theoretically ideal and can be reduced with poor instrumental resolution, e.g. significant axial divergence. The intersection of the change in slope from these different rates of decline can give an indication of the defect radius, R. This is simply the inverse of the distance in reciprocal space, i.e. $R \sim 1/(\Delta S)$, where the radius is defined as the transition of the strong to weak deformation. Thermal vibration falls into the category of weak modulation and will fall as $1/(\Delta S)^2$; this is commonly termed thermal diffuse scattering, TDS. We have only considered the diffuse scattering projected onto the scattering vector and clearly for a full analysis of the distribution of the diffuse scattering we should look in many directions (most conveniently with reciprocal space mapping, section 3.8.3) to evaluate the strain fields created by defects. Dederichs (1971) has considered some possible defects and how they influence the diffuse scattering. Since this diffuse scattering relates to the correlated movement of atoms detailed analysis have been possible for deriving the elastic properties and Debye temperatures of solids, etc.

2.8. Optical theory applied to reflectometry

In the discussion on dynamical theory we showed that when one wave enters the sample and no scattered wave is produced the incident beam is refracted and a specular reflected wave is produced, equation 2.33 and figure 2.9. One advantage of the dispersion construction, figure 2.9 and

2.10 is that it becomes clear that each intersection of the surface normal with the dispersion surface should create a contribution. If we refer to figure 2.10 we can see that there are several intersections in the two-wave dynamical model. These arise from the normal wave-fields described in the dynamical model and a specular wave. The specular wave will therefore exist over all scattering angles but is generally very weak and insignificant except close to the condition of grazing incidence. We shall now derive the form of this specular reflectivity.

Equation 2.33 indicates the effective refractive index for material is given by

$$n = (1+\chi_0)^{\frac{1}{2}} \sim \left(1 + \frac{\chi_0}{2}\right)$$
2.114

Where

$$\chi_0 = -\frac{e^2\lambda^2}{\pi m c^2}\rho_e = -\frac{r_e\lambda^2 Z}{\pi m_p A}\rho = -\frac{r_e\lambda^2 N_A Z}{\pi A}\rho = -\frac{r_e\lambda^2}{\pi V}F_0$$
2.115

from equations 2.20, 2.62 and 2.63. The electron density ρ_e has been replaced by the macroscopic density, ρ by transforming with the number of electrons in a unit volume, Z, the atomic mass of these atoms, A and the mass of the basic atomic building block, m_p of the hydrogen atom or Avogadro's constant N_A. This assumes that an average refractive index can be used is this case. At very low angles of incidence the X-rays are probing length-scales normal to the surface plane that are very large, i.e. $D \sim \lambda/2\sin\omega$ from Bragg's equation, well in excess of any interatomic distance, d. Hence this assumption is valid since we are averaging the spatially varying electron density, $\rho(r)$, over these long length-scales and this equates to the macroscopic average electron density, ρ, and similarly χ_0 is constant. Using a similar argument the absorption term must be some macroscopic average, which is included naturally as the imaginary component of F_0, whereas for the density it has to be included more explicitly. The phase of a wave at a position r along the wave-vector K into the sample will be given by

$$\Phi = \exp\{-2\pi i n K r\} = \exp\frac{\left(-2\pi i \{1-\delta-i\beta\}r\right)}{\lambda} = \exp\frac{-2\pi i\{1-\delta\}r}{\lambda}\exp\frac{-2\pi\beta r}{\lambda}$$

2.116

where $n = 1 - \delta - i\beta$, the refractive index arranged into real and imaginary components. δ is the real part of χ_0 in this case. As before we take the real part to contain the phase front and the imaginary part to include the absorption. Now the loss of intensity due to the linear absorption coefficient is given in equation 2.97 and this is the same as the imaginary term of equation 2.116, and hence

$$\beta = \frac{\mu\lambda}{4\pi}$$

2.117

The extra factor of two arises from the equation 2.116 being associated with amplitude and equation 2.97 being associated with intensity.

The reflection coefficient for X-rays from the interface between two materials is related to the difference in the wave-vector normal to the surface. The parallel component, as described earlier is unchanged, section 2.3.1. We can therefore write the reflection coefficient at an interface between two layers j and $j-1$ as

$$r_{j-1,j} = \left(\frac{K_{j-1} - K_j}{K_{j-1} + K_j}\right)_\perp$$

2.118

where K is the wave-vector normal to the surface.

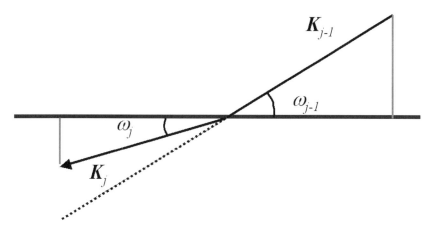

Figure 2.25 The influence of refraction on the incident beam to create a forward-diffracted beam.

Consider now figure 2.25 and include the conditions that we have stated

$$K_j \cos\omega_j = K_{j-1} \cos\omega_{j-1} = n_j k_0 \cos\omega_j = n_{j-1} k_0 \cos\omega_{j-1} \qquad 2.119$$

i.e. the parallel components of the wave-vector are equal and the wave-vector is changed in magnitude by the refractive index n compared with that in vacuum k_0, equation 2.33. Now equation 2.118 becomes

$$r_{j-1,j} = \frac{K_{j-1}\sin\omega_{j-1} - K_j \sin\omega_j}{K_{j-1}\sin\omega_{j-1} + K_j \sin\omega_j} = \frac{n_{j-1}\sin\omega_{j-1} - n_j \sin\omega_j}{n_{j-1}\sin\omega_{j-1} + n_j \sin\omega_j} \qquad 2.120$$

Now from equation 2.119

$$\cos\omega_j = \frac{n_{j-1}}{n_j}\cos\omega_{j-1} \qquad 2.121$$

and hence

$$\sin\omega_j = \left(1 - \left\{\frac{n_{j-1}}{n_j}\right\}^2 \cos^2\omega_{j-1}\right)^{\frac{1}{2}} \qquad 2.122$$

Therefore the reflection coefficient can now be determined in terms of the individual refractive indices and the incident angle on the interface concerned where there is a change in the refractive index.

$$r_{j-1,j} = \frac{n_{j-1}\sin\omega_{j-1} - (n_j^2 - n_{j-1}^2\cos^2\omega_{j-1})^{1/2}}{n_{j-1}\sin\omega_{j-1} + (n_j^2 - n_{j-1}^2\cos^2\omega_{j-1})^{1/2}} \qquad 2.123$$

We now have to consider the magnitude of the incident and reflected electric fields, D_0 and D_R respectively, which have the form given in equation 2.25.

$$\begin{aligned} &D_{R,m}\exp\{2\pi i S_m d_m/2\} \\ &D_{0,m}\exp\{-2\pi i S_m d_m/2\} \end{aligned} \qquad 2.124$$

This is the magnitude of the electric wave in the middle of the m th layer that has a thickness d_m. From figure 2.8 we can see that the change in the electric field component in the plane of the scattering (π polarisation) is insignificant for small incident angles whereas the electric field component parallel to the interface is almost entirely of the σ polarisation. We shall treat them equally. The wave-field components parallel to and at the interface (i.e. at a distance $d_{j-1}/2$ from the centre of layer $j-1$ and $d_j/2$ from the centre of layer j) are equal, therefore

$$(D_{0,j-1}e^{2\pi i S_{j-1}d_{j-1}/2} + D_{R,j-1}e^{-2\pi i S_{j-1}d_{j-1}/2}) = (D_{0,j}e^{-2\pi i S_j d_j/2} + D_{R,j}e^{2\pi i S_j d_j/2}) \qquad 2.125$$

As discussed above the reflectivity is given by the change in wave-vector normal to the surface plane. The component of the amplitude of the wave normal to the interface will be modified in proportion to the scattering vector. The scattering vector $S_j = 2K_j\sin\omega_j = 2k_0 n_j \sin\omega_j$, (i.e. twice the vertical component of the wave-vector) and since this is purely a specular reflection the scattering vector is always normal to the surface plane. For this to be continuous across the interface

$$(D_{0,j-1}e^{2\pi i S_{j-1}d_{j-1}/2} - D_{R,j-1}e^{-2\pi i S_{j-1}d_{j-1}/2})S_{j-1} = (D_{0,j}e^{-2\pi i S_j d_j/2} - D_{R,j}e^{2\pi i S_j d_j/2})S_j$$
2.126

Hence for a σ, π polarised or a circularly polarised incident beam the magnitude of these electric field amplitudes are in proportion, equation 2.125 and 2.126, and we can solve these two equations to give

$$X_{j-1} = (e^{-2\pi i S_{j-1}d_{j-1}/2})^4 \left\{ \frac{r_{j-1,j} + X_j}{r_{j-1,j}X_j + 1} \right\}$$
2.127

where

$$X_j = \frac{D_{R,j}}{D_{0,j}}(e^{-2\pi i S_j d_j/2})^2$$
2.128

and

$$r_{j-1,j} = \frac{S_{j-1} - S_j}{S_{j-1} + S_j} = \left(\frac{K_{j-1} - K_j}{K_{j-1} + K_j}\right)_\perp$$
2.129

Equation 2.128 is similar to our previous definitions, except that we have a phase term since we require the amplitude at the interface not at the midpoint of the layer j.

We now have a recursive expression for the reflected amplitude, X_{j-1}. Hence if we know the amplitude in the layer j, X_j (equation 2.128) and the reflection coefficient at the interface between the $j-1$ and the j th layers then we can determine the amplitude in layer $j-1$ (equation 2.127). This is where we now need to apply a boundary condition since we do not know the amplitude in layer j. We can assume that the amplitude at the bottom of the sample is zero, i.e. there is no intensity scattered upwards from the bottom surface, then $X_N = 0$, therefore we can determine the amplitude immediately above this first interface, X_{N-1} from equation 2.127. The reflection coefficient at each interface has to be determined first, equation 2.123, and this is also recursive since we only know the angle of incidence at the surface, ω_0.

The reflected intensity or reflectivity of the sample is given by

$$R_0 = \frac{I}{I_0} = X_0^* X_0 \qquad 2.130$$

I_0 is the incident beam intensity and I the specular reflected beam intensity. The suffix 0 refers to the air (~ vacuum) layer above the sample.

It is clear to see that there are many assumptions that can be made, but without a full investigation it is difficult to assess their impact. Consequently the work described here will use the derived equation from the above approach.

2.8.1. Some general conclusions from this analysis

Consider equation 2.115, where we have two ways of describing the susceptibility. Either we can consider the value as a function of the electron density or in terms of a structure factor. The structure factor for $H = 0$ is simply the sum of the electrons in the unit cell repeat and this is only applicable to crystalline solids. However the electron density can be related to the macroscopic density through Avogadro's number and the average atomic mass, equation 2.115, therefore we can apply this model to amorphous and poorly crystalline material. It is very important though to realise that amorphous and polycrystalline materials may not have the same density as their crystalline equivalents.

To obtain an indication of the form that this reflectivity takes, consider equation 2.123, for the case of a single homogeneous sample. We are now only interested in a single interface, the surface, hence

$$r_{0,1} = \frac{\sin\omega_0 - (n_1^2 - \cos^2\omega_0)^{1/2}}{\sin\omega_0 + (n_1^2 - \cos^2\omega_0)^{1/2}} \qquad 2.131$$

n_0 is of course unity (the refractive index of air) and n_1 is complex. However for incident beam angles for which

$$\omega_0 < \cos^{-1}\{\text{Re}[n_1]\} \qquad 2.132$$

the reflectivity is close to unity, although there is a small imaginary component that represents absorption into the sample. This penetration depth is very small. *Re* represents the real part of the quantity. When the angle of incidence reaches a value such that

$$\omega_0 > \cos^{-1}(\text{Re}[n_1]) \qquad 2.133$$

then the reflectivity begins to fall in value very rapidly. An indication of the shape is given in figure 2.26, as well as the variation in the penetration depth with incident angle.

As will be shown later in the examples, Chapter 4, the reflectivity from layer structures will exhibit oscillations characteristic of the thicknesses involved. In this case the penetration of the X-rays varies and is a maximum in the troughs of the oscillations and at a minimum at the peaks. The critical angle, when equations 2.132 and 2.133 become equalities, is clearly just a function of the material and the wavelength of the X-rays. From equations 2.114 and 2.115, we have

$$(\omega_0)_c \approx \cos^{-1}\left\{\text{Re}\left(1 - \frac{r_e\lambda^2}{2\pi V}F_0\right)\right\} \approx \cos^{-1}\left\{\text{Re}\left(1 - \frac{r_e\lambda^2 N_A Z\rho}{2\pi A} - i\frac{\lambda\mu}{4\pi}\right)\right\} \qquad 2.134$$

Also the derivation so far is based on an infinitely long sample that is homogeneous and abrupt at all its interfaces.

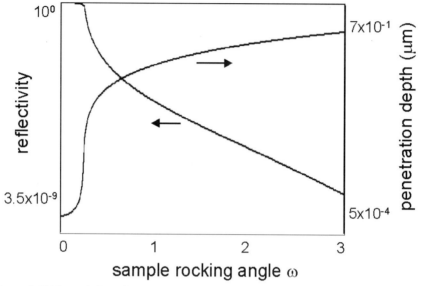

Figure 2.26 The variation of specular reflecting power of Si and its consequential penetration depth below the surface.

2.8.2. Imperfect interfaces

We can include imperfect interfaces in a number of ways. The simplest approach is to include a Gaussian smearing function to the reflectivity at each interface. The most robust method is to include many layers of different electron density close to the interface, just as in the case of the dynamical model of section 2.4. We can now calculate the influence of roughness on the shape of the specular profile for a single layer on a substrate, by varying the spreading of the surface and interface electron densities, figure 2.27.

The description of interfacial roughness considered above is based on the averaging of the density parallel to the interface and this is applicable for interfaces dominated by interdiffusion and segregation. For laterally rough (jagged) interfaces we need to consider a different approach.

Chapter 2 An Introduction to X-Ray Scattering 91

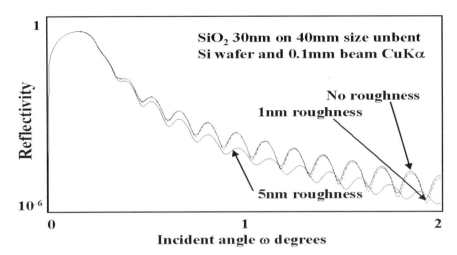

Figure 2.27 The simulated profile of Si with a SiO$_2$ layer with varying degrees of roughness. Note how the region of "total" external reflection is altered from the simulation of figure 25, because of the inclusion of finite sample and beam size. The fringe period increases as the roughness takes up more of the layer.

We may understand the process by understanding that the reflectivity is based on the local normal of the interface. The reflectivity will therefore be changed with local density undulations not parallel to the interface. Considering the process of scattering in this way we immediately see the complications: multiple reflections can take place as well as transmission, and hence refraction, through jagged regions. Also the scattering will almost certainly be out of the general plane of scattering. However a number of authors have taken a more pragmatic approach by assuming that finite correlation lengths obtained from the diffuse scatter profiles can be directly attributed to a measure of lateral roughness. For two regions of different density separated by a distance *r* we can define some scattering function as the Fourier transform of a pair correlation function, $g(r)$

$$G(S) \propto \int g(r) \exp^{2\pi i S \cdot r} dr \qquad 2.135$$

The resultant diffuse scattering is then a function of the incident wave intensity, this scattering function and transmission coefficients (i.e. the modification to the incident and scattered intensity because of the depth of the interface below the surface, this s the basis of the Distorted Born Wave Approximation, DWBA).

$$I_{diffuse}(S) = I_0 \mid t(k_0) \mid^2 \mid t(k_H) \mid^2 G(S) \qquad 2.136$$

$t(k_0)$ and $t(k_H)$ are the transmission coefficients of the incident beam and scattered beam respectively for the scattering geometry to create S; also $t = 1 - r$, equation 2.129. Hence from knowledge of the pair correlation function $g(r)$, we can determine the distribution of diffuse scattering associated with roughness. This function can be considered in its most general form as the correlation of some density at height z_1 normal to the interface at position x_1 with that at another height z_2 and position x_2

$$g(X) = <z(x)z(x+X)> \qquad 2.137$$

For a single interface the following relationship can be assumed to represent a good approximation, Sinha et al (1988)

$$g(X) = \sigma^2 \exp\{-\left(\frac{\mid X \mid}{\xi}\right)^{2h}\} \qquad 2.138$$

where σ is the root mean square roughness (r.m.s.) and ξ is the lateral correlation length. The parameter h is an additional variable that has a value between 0 and 1 and can be related to the fractal parameter D ($= 3 - h$). Following the derivation of Sinha et al we can determine the scattering function for his case as

$$G(S) = \frac{e^{-4\pi^2 \mid S_\perp \mid^2 \sigma^2}}{4\pi^2 \mid S_\perp \mid} \int \{e^{-4\pi^2 \mid S_\perp \mid g(X)} - 1\} e^{-2\pi i S_\parallel X} dX \qquad 2.139$$

S is resolved into components perpendicular and parallel to the interface. To help understand the general shape of this function, let us first assume that the pair correlation parameter $g(X)$ to be small, and since $S_\perp$ is small, $S_\perp g(X) << 1$ then

$$G(S) \approx e^{-4\pi^2 |S_\perp|^2 \sigma^2} \int g(X) e^{2\pi i S_{//} g(X)} dX \qquad 2.140$$

Therefore when $h \sim 1$ the scattering function $G(S)$ is Gaussian in shape and when $h \sim 0.5$, $G(S)$ is Lorentzian shape. For $h \sim 0$ the pair correlation parameter is constant and represents a constant scattering function characteristic of very rough or jagged interfaces. A profile based on this is illustrated in figure 2.28.

We can use these expressions for roughness provided that the roughness is uncorrelated from layer to layer by considering each interface to contribute independently. The roughness contribution is therefore the sum of these contributions after taking into account transmission coefficients. The correlation parameter to take account of the roughness replication from layer to layer in approximate form can be expressed by

$$g_{j,k}(\mathbf{x},z) = \frac{1}{2}\left(\frac{\sigma_k}{\sigma_j}g_j(\mathbf{x}) + \frac{\sigma_j}{\sigma_k}g_k(\mathbf{x})\right) e^{-\frac{|z_j - z_k|}{\xi_v}} \qquad 2.141$$

where j and k are the layer indices and ξ_v is the vertical correlation length, Ming, Krol, Soo, Kao, Park and Wang (1993).

Now this expression is only valid when all lateral correlation lengths are correlated in depth through the structure, which will be an assumption. It is easier to envisage vertical correlation of long lateral correlation lengths, whereas the short lateral correlation lengths (high frequency components) are less likely to be replicated. This can be incorporated by introducing an x^n dependence in the exponent.

It is very important to note that interfaces are not necessarily fractal and examples of other interfaces are illustrated in figure 2.28 and 2.29.

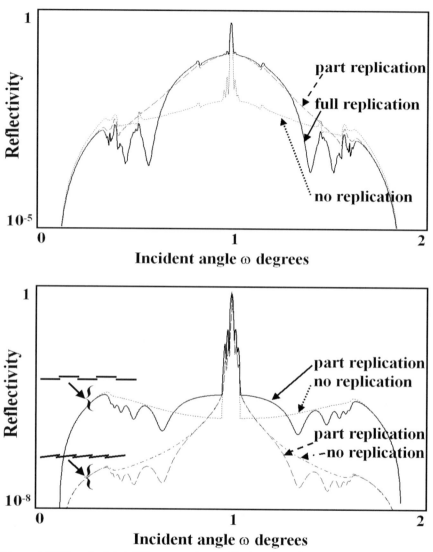

Figure 2.28 The calculated diffuse scattering profile for a laterally rough interface from a periodic GaAs/AlAs (15nm;7nm) multiplayer on GaAs, assuming a fractal model of the roughness of 3 monolayers, (top), assuming a half monolayer coverage on a flat surface (terrace width of 164nm) and monolayer steps created by a 0.1^0 inclined surface (terrace width of 164nm) (lower).

Chapter 2 An Introduction to X-Ray Scattering 95

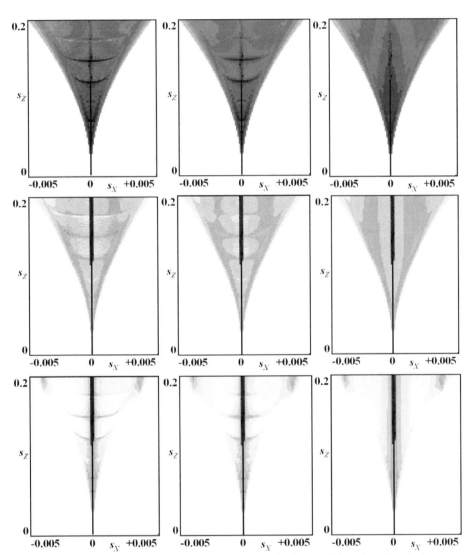

Figure 2.29 The calculated diffuse scattering patterns for the conditions described in the caption to figure 2.28, i.e. for fully correlated, partially correlated and not correlated roughness (left to right). (top) fractal model plotted over 6 orders in intensity, (middle) flat surface half-monolayer coverage plotted over 9 orders of intensity and (bottom) a monolayer stepped surface resulting from a 0.1^0 <001> inclination to the surface normal. The observed lateral streaking is characteristic of vertical correlation and the specular scan is at $s_x=0$.

A fractal interface cannot be the correct model for a nearly perfect interface, since the dimensions at the atomic level are discrete and therefore as an example a GaAs (001) surface must represent a minimum step height of a monolayer. Hence the interface roughness for the fractal model in figures 2.28 and 2.29 are for a three monolayers, to give an approximate feasible fractal form. An almost perfect interface can be represented as a series of single monolayer steps distributed over the surface resulting from incomplete coverage. The diffuse scattering from this is different from that from a fractal surface and this becomes more pronounced as the replication from layer to layer becomes weaker. For a vicinal surface the correlated roughness is inclined at an angle of φ, this will have the correlated roughening of layer j displaced from layer k by a distance $(z_j - z_k)\tan\varphi$ therefore x becomes $x-(z_j - z_k)\tan\varphi$, etc., Kondrashkina, Stepanov, Opitz, Schmindbauer, and Kohler (1997). Again the effect can be quite dramatic, figure 2.28 and 2.29, indicating the sensitivity of the diffuse scattering to the interface model, Holy and Fewster (2002).

We immediately see that the complexity is growing rapidly and the number of parameters that should be refined are more uncertain, however this is where the instrument resolution or scan range can limit the spread of correlation lengths relevant to the calculation. As the resolution leads to smearing of the features the longer length scale limit is reduced and the possible scan range puts a lower limit on the correlation length measurable. Let us consider some approximate values. For CuKα radiation and a divergence of the incident beam of $0.01°$ then when the sample is rocked about a scattering angle of $1°$ the lateral correlation lengths measurable range from 4.4nm to 440nm at best.

2.9. In-plane scattering

So far we have considered several theoretical models that will cope with the most perfect crystalline materials to those that are independent of crystalline form. These are all applicable to the types of analytical problems in semiconductors. We can also combine our understanding of the various scattering models in the following way. Suppose we have an X-ray beam incident at a very low angle, then the penetration is very small, figure 2.26. However because there is some absorption some energy penetrates and is available for scattering. Scattering planes that can easily be brought into the correct angle for scattering are those roughly perpendicular to the sample surface normal, figure 2.30. Marra, Eisenberger and Cho (1979) first exploited this approach.

We can consider the scattering in a similar way to those approaches considered above but our boundary conditions in the case of the dynamical model are changed. The penetration depth can be calculated and this is a useful variable in the experiment. The scattering will also not exit parallel to the surface except under special conditions and will depend on the incident beam angle.

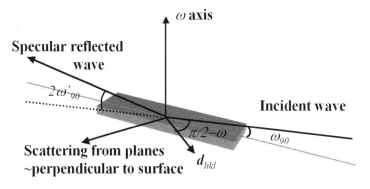

Figure 2.30 The geometry of the in-plane scattering experiment.

Suppose we now have an incident wave making an angle to the surface of ω_{90} and an angle ω to the scattering plane then its component along the scattering vector S is $k_0 \cos\omega_{90}\sin\omega$. The component of the resulting scattered wave will similarly be $k_H\cos(2\omega_{90}'-\omega_{90})\sin(2\omega'-\omega)$, where $(2\omega_{90}'-\omega_{90})$ is the exit angle we wish to know. Using the Bragg condition we obtain

$$k_H \cos(2\omega'_{90}-\omega_{90})\sin(2\omega'-\omega) = -k_0 \cos\omega_{90}\sin\omega + S \qquad 2.142$$

and since $S = 2\sin\theta/\lambda\{1+\chi_0/2\}$ and $|k_H| = |k_0| = 1/\lambda$ then

$$(2\omega'_{90} - \omega_{90}) = \cos^{-1}\left\{\frac{2\sin\theta\sin\varphi\left[1+\frac{\chi_0}{2}\right] - \cos\omega_{90}\sin\omega}{\sin(2\omega'-\omega)}\right\} \qquad 2.143$$

where φ is the angle that the scattering plane makes with the surface. Now $\omega = \omega' = \theta$, since the concept of scattering planes inclined to the scattering plane is no longer relevant then

$$(2\omega'_{90} - \omega_{90}) = \cos^{-1}\left\{2\sin\varphi\left[1+\frac{\chi_0}{2}\right] - \cos\omega_{90}\right\} \qquad 2.144$$

So if we consider the case that $\varphi = 90^0$ then

$$(2\omega'_{90} - \omega_{90}) = \cos^{-1}\{[2+\chi_0] - \cos\omega_{90}\}$$

therefore using the expansion for small angles $\cos\omega_{90} \sim 1 - \omega_{90}^2/2$, etc., and simplifying we obtain

$$\cos(2\omega'_{90} - \omega_{90}) = \cos(\omega_{90}) \sim 1 - \frac{\omega_{90}^2}{2} \sim 1 + \chi_0 + \frac{\omega_{90}^2}{2}$$

For the scattered wave to leave the sample through the entrance surface this angle must be positive and real and consequently

$$\omega_{90} < \{-\chi_0\}^{1/2}$$

for this condition. Comparing this with equation 2.132 and using equation 2.114 we find that

$$\omega_{90} < (\omega_0)_c \qquad 2.145$$

That is, to a first approximation the incident angle must always be less than the critical angle to obtain an exit beam from the entrance surface. A method of simulating the scattering from structures with rough interfaces has been given by Stepanov, Krondrashkina, Schmidbauer, Kohler, Pfeiffer, Jach and Souvorov (1996). The method is based on dynamical theory and accounts for all the scattering contributions; effectively bringing together the optical and dynamical components discussed in this chapter.

This method of collecting data can prove very useful since it gives access to scattering from planes that are normally only accessible with transmission geometry.

2.10. Transmission geometry

This will be covered very briefly here for completeness although much has been discussed previously on this configuration because of its importance in transmission topography, Authier (1996), Tanner (1996). The theoretical basis is no different from that described above except the boundary conditions are different and therefore the intersection of the surface normal in figure 2.10 will occur on both branches of the dispersion surface. The excitation of additional wave-fields creates fascinating interference effects that have been used to vindicate the assumptions in dynamical theory. The most extensive use of transmission geometry (or Laue geometry) has been for high-resolution topography of Si wafers for imaging defects created during processing. Since in general simulation of the contrast is less of a concern, whereas direct interpretation of the contrast is of interest, we will take a very pragmatic approach in this section.

From the Takagi theory, section 2.4, it was clear that the displacement field will be influenced but is still maintained close to small distortions. Therefore the path of the X-rays will be deviated thus changing the distribution of intensity emerging from the sample. If the distortion is too large as in the case close to a dislocation then the wave is uncoupled and will be scattered out as described in section 2.5. Of course the strain close to a defect and the associated crystal plane curvature diminishes with distance from the defect centre. However because of the high sensitivity of X-rays scattering to small strains and small curvatures the image is large and therefore clearly visible on a direct image with no magnification. We shall consider a few cases to give an outline of how the contrast is created in samples of various thicknesses. The important definition that determines the likely contrast is the product of the thickness, t and the linear absorption coefficient, μ, equation 2.97, since this influences the X-ray penetration depth. The basis of X-ray topography, which is our only concern in this section, is that the scattered image has a finite size and we wish to examine the contrast within this scattered diffraction spot. The contrast in this spot image approximately relates to the region of the sample that is probed by the incident X-rays.

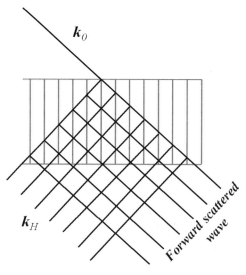

Figure 2.31 The complex wave-field created in transmission geometry (cf. Figure 2.6). When the absorption increases this Borrmann triangle reduces and transmission is confined to the direction parallel to the scattering planes.

Suppose the sample is very thin or weakly absorbing, $\mu t \sim 1$. From the dynamical scattering theories discussed above we can see that the intrinsic scattering width for a perfect crystal is very small. If we now probe our sample with an incident beam that has a divergence larger than this, we have to visualise two regions; those that are perfect crystal and those that are not. The perfect region will scatter at the correct scattering angle, within a very narrow angular range, i.e. it will extract a very small proportion of the incident X-rays for scattering. However close to defects there are a range of angles that are presented to the divergent forward-refracted wave and thus there are many more possibilities for scattering outside the intrinsic scattering width of the perfect regions. The intensity from the defects is therefore enhanced with respect to the perfect regions, provided that the absorption is not too high. X-ray topographic images of defects closer to the exit surface of the sample are better defined than those close to the entrance surface. This occurs because the proportion of the Borrman triangle, figure 2.31 that creates the relative contrast is diminished and the scattered waves from the defects are partially absorbed.

As the sample thickness is increased such that $\mu t \sim 10$, the absorption appears to be so large that no X-rays will pass through the sample. However this average linear absorption coefficient is only a factor used in the kinematical theory, section 2.7.2 and applied in the optical theory, section 2.8. The dynamical model on the other hand makes no such assumption and in fact for the transmission case four tie points can be excited (both branches of the dispersion surface for both polarisation states) and each will have their own characteristic absorption. From section 2.3, we established that the imaginary component of the wave-vector susceptibility relates directly to the absorption and we have derived an expression relating the linear absorption coefficient, equation 2.117. If we compare this with equation 2.115 then

$$\mu = -4\pi \, \text{Im}(\chi_0 \mathbf{K}_m) \qquad 2.146$$

where $\text{Im}(\chi_0 K_m)$ represents the imaginary component of $\chi_0 K_m$, $|K_m|=1/\lambda$. The attenuation of interest is in the direction of energy flow into the crystal, whereas equation 2.146 represents the influence normal to the surface, figure 2.7. We can see from figure 2.10, that if several points are excited on the dispersion surface (i.e. on branch 1 and branch 2) then the internal wave-vectors will have different components along the surface normal and consequently different linear absorption coefficients. Similarly the polarisation factor creates a further offset of the dispersion surfaces and therefore the four existing waves have their own linear absorption coefficients. Bonse (1964) gives the full expression.

$$\mu_w = \mu \frac{\left\{ 1+|X|^2 + 2\left[\frac{C \, \text{Re}(_w\chi_H X)}{\text{Im}(_w\chi_0)} + \text{Im}(_w\chi_H X) \right] \right\}}{\left\{ 1+|X|^4 + 2|X|^2 \cos 2\theta \right\}^{1/2}} \qquad 2.147$$

where the subscript w is the index for the appropriate wave-field and the other symbols have been defined previously. What is found from using this formula is that the linear absorption coefficient for different wave-fields varies dramatically and thick crystals are still able to allow energy to flow. The physical explanation is that the wave amplitude oscillates with the periodicity of the lattice such that the nodes correspond to the atomic sites and the absorption is low. There is of course a similar wave that has its anti-nodes at the atomic sites and is soon absorbed in a thick crystal. The energy flow for this low absorbing wave through the crystal is parallel to the atomic planes.

Although this anomalously transmitted wave is rather weak it is a very powerful probe for imaging defects. Clearly any defect close to the entrance surface will scatter this wave and the intensity will be lost, giving rise to missing intensity in the image. If the degree of perfection is good until close to the exit surface then the curvature of the planes around a defect will "steer" the energy flow. This will give rise to enhanced intensity when it is directed more to the scattered wave direction and reduced intensity when directed more towards the incident beam intensity, giving rise to black-white contrast. In general these principles can be used in the interpretation of most X-ray topographs taken in transmission geometry. Although a deep theoretical understanding is necessary for modelling images and greater depth is required for interpreting subtle dynamical scattering features, that is not our main concern here.

2.11. General conclusions

The theoretical models described here cover many aspects of the scattering that we wish to interpret from semiconductor materials. The models discussed are generally applicable but the reader should be aware that we have largely concentrated on the interaction of a single photon with an initial pre-defined path. In general any source is divergent and we should consider the wave as having curved wave-fronts, Kato (1968a and b), although the Takagi theory partially overcomes this assumption. Kato approached this by making a superposition of plane waves and this is the basis of the approach used in the next Chapter to take into account divergence in the X-ray source due to the instrumentation.

We have also not considered partial coherence in this Chapter due to wavelength spread, again we take a rather pragmatic approach and include it in the dynamical modelling of the scattering, i.e. add it as an instrumental effect, Chapter 3. However there are considerable possibilities in refinement, validating and extending as the materials challenge increases. As we shall see we cannot isolate all the components in an X-ray diffraction experiment and we have to combine the scattering theory, sample and data collection instrumentation as a whole. The next chapter will cover the instrumental aberrations.

References

Authier, A (1996) in *X-Ray and Neutron Dynamical Diffraction: Theory and Applications.* pp 1-31 and 43-62. Editors Authier et al, Plenum Press: New York.
Bonse, U (1964) Z Phys. **177** 385
Compton, A H (1917) Phys Rev 9 29
Darwin, C G (1914a) Phil. Mag. **27** 315
Darwin, C G (1914b) Phil. Mag. **27** 675
Debye, P (1914) Ann. D Physik, **43**, 49
Debye, P (1915) Ann der Physk 46 809
Dederichs, P H (1971) Phys. Rev. B **4** 1041
Ewald, P P (1916a) Ann. d Physik **49** 117
Ewald, P P (1916b) Ann. d Physik **49** 1
Ewald, P P (1917) Ann. d Physik **54** 519
Fewster, P F (1986) Philips J Res. **41** 268
Fewster, P F (1992) J Appl. Cryst. **25** 714
Hartree, D R (1935) Proc. Roy. Soc. **A 143** 506
Holy, V (1982) Phys. Stat. Solidi B **111** 341
Holy, V and Fewster, P F (2002) unpublished collaborative work, also see Holy, V, Pietsch, U and Baumbach, T (1999) *High Resolution X-Ray Scattering* Springer: ISBN 3-540-62029-X
Holy, V and Fewster, P F (2003) J Phys. D to be published
Huang, K (1947) Proc. Roy. Soc. A **190** 102
Fock, V (1930) Zeit. F Physik **61** 126
Kato, N (1968) J Appl. Phys. **39** 2225
Kato, N (1968) J Appl. Phys. **39** 2231

Kato, N (1980) Acta Cryst. **A36** 763

Khrupa, V I (1992) Phys. Metals **11** 765

Kondrashkina, E A, Stepanov, S A, Opitz, R, Schmindbauer, M and Kohler R (1997) Phys. Rev. B **56** 10459

Krivolglaz, M A (1995) *Theory of X-ray and Thermal Neutron Scattering by Real Crystals* (New York: Plenum)

Laue M v (1931) Ergeb. Der exact. Naturwiss **10** 133

Marra, W C, Eisenberger, P and Cho A Y (1979) J Appl. Phys. **50** 6927

Ming, Z H, Krol, A, Soo, Y L, Kao, Y H, Park, J S and Wang, K L (1993) Phys Rev B **47** 16373.

Olekhnovich, N M and Olekhnovich, A I (1981) Phys. Stat. Solidi A **67** 427

Pavlov, K M and Punegov, V I (1998) Acta Cryst. **A54** 214

Prins, J A (1930) Zeit. F Physik **63** 477

Reid, J S (1983) Acta Cryst. A39 1

Scherrer, P (1918) Gott. Nachr **2** 98

Sinha, S K, Sirota, E B, Garoff, S and Stanley, H R (1988) Phys. Rev. B **38** 2297

Stepanov, S A, Krondrashkina, E A, Schmidbauer, M, Kohler, R, Pfeiffer, J-U, Jach, T and Souvorov, A Yu (1996) Phys. Rev. B **54** 8150.

Takagi, S (1969) J Phys. Soc. Japan **26** 1239

Takagi, S (1962) Acta Cryst. **15** 1311

Tanner (1996) in *X-Ray and Neutron Dynamical Diffraction: Theory and Applications.* pp 147-166. Editors Authier et al, Plenum Press: New York.

Taupin, D (1964) Bull. Soc. Fran. Miner. Cryst. **87** 469

Zaus, R (1993) J Appl. Cryst. **26** 801

Chapter 3

EQUIPMENT FOR MEASURING DIFFRACTION PATTERNS

3.1. General considerations

The instruments to obtain a diffraction pattern consist of a few basic features: an X-ray source, incident beam conditioning, sample stage and scattered beam capture. The nature of these features should be selected to best meet the needs of the material property to be analysed.

In achieving the ideal combination to analyse the material we have to balance diffraction pattern resolution, real space resolution, intensity and data collection time. These parameters are best understood by having a good grasp of the physics associated with each of the incident and scattered beam conditioners, as well as the X-ray source and detector electronics. Since we are primarily interested in extracting certain material properties from a diffraction pattern it is important to know all those features that can influence this pattern. Clearly if we know little about the sample and wish to extract a lot of information then the instrument function has to be very well understood. If we already know a lot about the sample then perhaps we can perform a simpler experiment with a less well-defined instrument function. One very good feature concerning semiconductors is that we generally know a reasonable amount of information (approximate thickness and composition in each layer) and because they are amenable to a well-defined instrument function we can extract a real wealth of additional information.

A modern X-ray diffractometer is illustrated in figure 3.1, where the incident beam and scattered beam optics can be exchanged to suit the material problem to be analysed. The sample stage in this case is a precision goniometer with optical encoders on the axes, thus eliminating angular uncertainties on the ω and $2\omega'$ main axes, figure 3.2. The χ, tilt axis, allows for 180° rotation and the ϕ rotation axis normal to the sample surface can rotate through 360°. The sample can also be positioned with an *xyz* stage. The $2\omega'$ angle can have a defined zero angle related to the direction of the incident beam, whereas the ω angle can be conveniently defined relative to the sample surface, if it is flat. With an instrument of this type the sample can be manipulated into any position and the incident beam and scattered beam optics chosen to suit the materials problem.

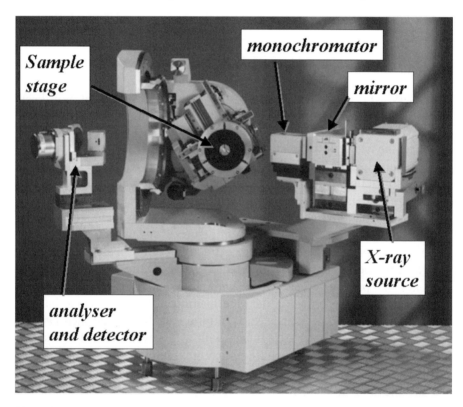

Figure 3.1. A modern versatile diffractometer from PANalytical. The X-ray source, monochromator, mirror and analyser assemblies are all pre-aligned and can be exchanged for alternative components. This allows the matching of the instrument to the sample problem to be analysed.

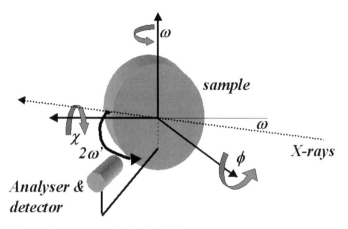

Figure 3.2. The angles associated with the diffractometer movements.

The development of optical elements for diffractometers is happening very fast and the number of options is increasing significantly. With the instrument described above the various modules can be added resulting in an ever increasingly versatile base. The present day commercial instrumentation is becoming increasingly sophisticated as the demand rises, but also for ease of use and to meet strict safety requirements and tolerances. Therefore just as in the revolution in electronics the building blocks are getting bigger and the need for homebuilt instrumentation of this kind is reducing rapidly.

We have discussed the scattering processes in Chapter 2 and now we will consider the incident beam conditioning and scattered beam capture. As with all these approaches they fall into two main components, active and passive components. Active components consist of crystals or absorbers that can dramatically change the spectral characteristics and passive components consist of slits and such like that limit the beam that can pass. Firstly we will consider the basics of the resolution on the capture region of the diffraction pattern and the main components of the X-ray source and some detectors.

3.2. Basics of the resolution function

From the geometry of the dispersion surface, figure 2.10 or the basic relation $k_H = k_0 + S$, equation 2.13 we can draw the condition of scattering, figure 3.3. We have restricted the case to reflection from crystals, although to include all cases of reflection and transmission the accessible region occupies a full sphere. The reciprocal lattice for the sample of interest is represented by dots and this exists in reality as a three dimensional array. A reciprocal lattice point is at a distance $1/d_{hkl}$ from the origin O and therefore can give rise to scattering, section 2.2.2 (equation 2.8). The position of each reciprocal lattice point represents the inverse of the interplanar spacing, defined by the distance from the origin, and the direction of the plane normal. Figure 3.3 is illustrated for a perfect single crystal. For a composite structure, e.g. a semiconductor multilayer, we can superimpose several reciprocal lattice meshes and this provides a good way to visualise the region of reciprocal space to investigate. A few examples are given in figure 3.4. From this figure we can relate our reciprocal space co-ordinates of our scattering vector; q_x and q_z or s_x and s_z to the incident angle ω and scattering angle $2\omega'$

$$s_x = \frac{q_x}{2\pi} = \frac{1}{\lambda}\{\cos\omega - \cos(2\omega'-\omega)\}$$
$$s_z = \frac{q_z}{2\pi} = \frac{1}{\lambda}\{\sin\omega + \sin(2\omega'-\omega)\}$$
3.1

The reciprocal space co-ordinates s_x and s_z are on the same scale as wave-vectors and scattering vector S whereas the vector Q is equivalent to $2\pi k$. Throughout this book we have concentrated on S and k because this gives a reciprocal lattice constructed as the inverse of the interplanar spacings. We can also then simply see that if we change the wavelength the region of capture is changed.

The shape of the reciprocal lattice point contains information about the internal arrangement of the atoms in the structure and the distribution and size of the regions having this interplanar spacing. We can therefore consider each point to include contributions from internal strains, angular misorientations, finite size effects and their distributions. In fact all the structural details with varying degrees of each contribution. Immediately we see the reciprocal lattice point has a three-dimensional form and contains a considerable amount of information. For determining the average structure, as in conventional molecular structure determination, the concern is to capture the total intensity value from each reciprocal lattice point. However our main concern here is any deviation from the average structure, whether this is a composite layer structure of similar materials or details of the microstructure, which manifests as intensity variations within the reciprocal lattice spot. Hence our requirements for the reciprocal space resolution is very different from that used for determining the average structure.

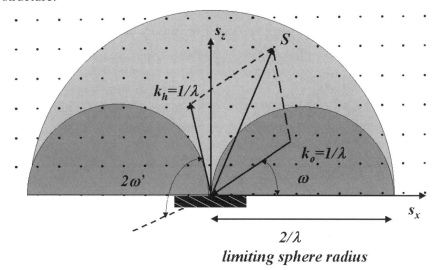

limiting sphere radius

Figure 3.3. The region of reciprocal space that can be captured in the reflection geometry (pale grey) and that which is inaccessible, except in special circumstances (dark grey). The incident and scattered beam wave-vectors indicate the angles of the diffractometer to capture the intensity associated with the scattering vector.

In future sections we will discuss the divergence from various X-ray collimators and in this section we will briefly explain how this defines the region of reciprocal space that we sample. From figure 3.5, we have considered an incident wave vector k_0 on the sample surface and allowed the incident angle ω to vary over a range of divergence $\Delta\omega$ and similarly we have placed our detecting system to receive scattered X-rays over a finite angular range. These finite angular divergences create a finite size probe on

110 X-RAY SCATTERING FROM SEMICONDUCTORS

our reciprocal space mesh. The X-ray wavelength is also not single valued and hence our incident and scattered beam vectors will also have a spread of lengths. Since we are only concerned with coherent elastic scattering the two vectors are the same length for each incident wavelength. The capture area has now enlarged and will vary throughout reciprocal space. As mentioned above the reciprocal lattice points do have a finite dimension in all directions, i.e. out of the plane of the figure as well as in the plane. Similarly there is a divergence (axial divergence) normal to the divergence in the scattering plane and therefore this captured region is a projection, but the measured intensity is assigned to the midpoint of the angle readout of the axes. Suppose now we wish to map the intensity distribution around a reciprocal lattice point then our probe size should be smaller or at least comparable with any intensity variations formed by the structural details of interest.

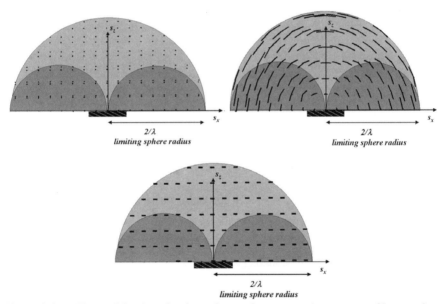

Figure 3.4. (a) The modification of reciprocal space for a composite structure of layer and substrate where the lattice parameter in the plane of the interface is the same. (b) An image of reciprocal space for a sample of several crystal orientations and (c) the case for a structure with a limited lateral dimension.

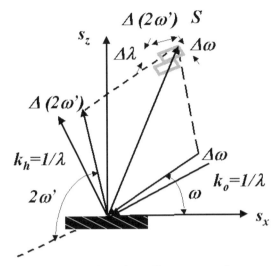

Figure 3.5. The influence of the incident beam divergence, analyser acceptance and wavelength spread on the region of capture of reciprocal space. The axial divergence is normal to the plane of the figure.

The majority of experiments are conducted with reasonable to good resolution in the scattering plane but poor resolution in the axial plane. This can lead to the wrong interpretation and will be discussed in an example of three-dimensional reciprocal space mapping, sections 3.8.3.3 and 4.3.2.3.

3.3. X-ray source

X-rays are generated by electron energy transitions in a solid (laboratory source) or in a confining magnetic field (synchrotron source). If the energy transition is well defined as in a solid (quantum transitions) then the energy of the emitted photon can be very well defined. X-rays are generated from transitions to the innermost electron orbitals and are characteristic of the atom concerned. The common laboratory sources are composed of copper, molybdenum, cobalt, iron and silver anodes for diffraction with most experiments favouring copper. The emission lines arise from excitations that transfer sufficient energy to remove an inner electron and allow the more loosely bound (higher energy states) to transfer to the vacant inner states. For example an electron in the L shell transferring to the K shell will create a Kα photon and an electron transferring from the M shell to the K will create a Kβ photon, etc. Of course there are more transitions than this and the Kα line is a doublet representing transitions between $2p^{3/2}$ to 1s (Kα$_1$) and $2p^{1/2}$ to 1s (Kα$_2$) states. The Kβ lines come from the 3p states and again the lines are not single valued. Basically we should consider that the intensity from an X-ray source is a complex function of wavelength that has been very closely calculated and is essentially invariant for any given tube current. The electrons that do not excite these lines will loose energy and emit a continuous background giving a total distribution of intensity with wavelength given in figure 3.6. The controls we have are the tube voltage and tube current and we should optimise these for the maximum intensity and stability. In general the intensity increases linearly with tube current and as the square with the voltage above the critical excitation voltage. However increasing the voltage too high will create some self-absorption, i.e. the X-rays are generated on average deeper into the anode and are partly absorbed. The simplest method of determining the optimum tube voltage and current is by experiment. For a 2kV X-ray Cu anode tube the typical values of 45kV and 40mA are close to optimum when considering all aspects of focus stability.

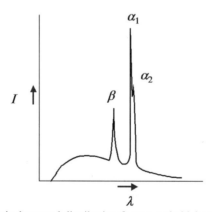

Figure 3.6. A typical spectral distribution from a sealed laboratory X-ray source.

The X-rays are generated in a confined area; generally termed the focus of the anode that is a function of the tube geometry. The thermal load from high-energy electrons impacting on the anode is very high and efficient cooling is essential in good X-ray tube design. A balance is therefore always necessary between focus size and the power available.

The filament, figure 1.4, is a small linear coil and its dimension partially defines the focus and this is approximately rectangular in shape. The advantage of this shape is that the projection normal and parallel to the long axis produces two very useful X-ray source shapes, figure 3.7. The projection angle can be changed a little giving rise to improvements in resolution or an increase in intensity. Again we have to compromise depending on the nature of the experimental problem to be resolved.

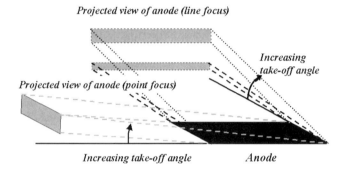

Figure 3.7. The different projections available from a sealed laboratory X-ray source.

3.4. X-ray detectors

This will only be covered briefly since in many respects this is a sealed component that is generally very reliable, however there are some basics that we should discuss. The ideal requirement of any detector is to record every X-ray photon and produce a measurable signal proportional to the number of photons arriving per second over a large flux range. To maintain the linearity in the response of X-ray photons arriving at the detector to the recorded signal we have to reduce the contribution of any residual noise for weak signals and compensate for any time when the detector is inactive due to high fluxes. Most detectors work on the basis of the X-ray photon creating electron-ion (or hole) pairs either in a gas or a solid.

3.4.1. The proportional detector

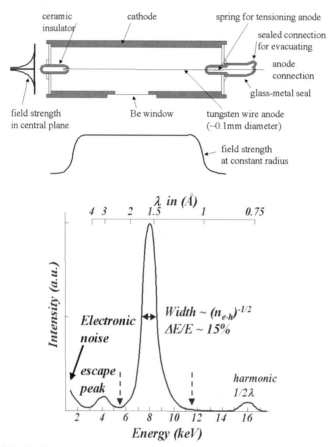

Figure 3.8. (a) The proportional X-ray detector. (b) The energy distribution for discriminating CuKα X-rays with a proportional detector.

One of the most reliable detectors is the proportional counter that is simply a sealed cylindrical outer electrode with an inner coaxial wire kept at a large positive voltage with respect to the outer electrode, figure 3.8a. The region between the electrodes is filled with an inert gas that acts as an insulator at high electric fields. To obtain a highly efficient counter we require all X-rays to be absorbed within the gas, this partially determines the choice of inert gas. Argon is a low absorber and therefore not ideal, Krypton favours radiative absorption processes and is hence unsuitable for creating significant current signal. Xe creates Auger electrons and gives about 93% and 45% absorption efficiencies for Cu and Mo X-rays respectively and with careful design can produce a high quality detector. The gas pressure can also influence the sensitivity to different energies and allows for some further optimisation. An X-ray transparent window (Be or mica) is placed at the side of the cylinder. X-rays entering the window then ionise the gas and the intense electric field accelerates the electrons towards the wire creating further impact ionisation events, more electron-ion pairs, and a recognisable current pulse that can be amplified and detected. This gas amplification, $\sim 10^4$, is important for creating a large signal compared with any residual noise. The pulse is further amplified with a low noise high quality linear amplifier mounted close to the counter to reduce the possibilities of additional noise. The highest electric field is very close to the wire and this is where most of the impact ionisation takes place. The electric field must be chosen carefully to give a strong pulse but not increased too much otherwise the plasma created will modify the local electric field significantly and the proportionality will be lost.

The energy of the X-ray photon determines the number of electron-hole pairs and therefore influences the strength of the current pulse in the detection circuit. The favoured absorption mechanism for reliable counting is the Auger process, which is non-radiative and ejects electrons. Xe is therefore the best choice of inert gas. The energy to eject an electron from a Xe atom is ~20.8eV, hence an 8.04KeV CuKα photon has enough energy to create ~386 electron-ion pairs. The residual energy from the first ionisation event is converted into the momentum of the ejected electron that in turn creates further ionisation. This aspect gives the energy discrimination. The additional energy from the electric field increases the number of electron-ion pairs further and this is the signal pulse of interest.

Suppose a proportion of the created ions return from their excited state by internal recombination with the emission of radiation. This fluorescent radiation created takes most of the energy and the ejected electron has a lower momentum and consequently creates a weaker pulse. This weak pulse is often termed the escape peak and corresponds to a low energy peak in the detector response function. Statistically this will be a peak of a fixed ratio with respect to that created by the Auger process. Leakage currents associated with these very high voltages are inevitable and will create a further signal, but just as with the escape peak, it will create a different pulse strength from that of the wanted signal. This signal will be fairly continuous and hopefully well below the pulse strength of a photon interaction with gas amplification. The detector response function is shown in figure 3.8b.

The most important aspect of any detector is the relationship between the signal and the incoming photon flux. The good energy resolution, figure 3.8(b) helps enormously in obtaining a stable and proportional response. Clearly though we should isolate the leakage current and the escape peak from the signal current and this is achieved with a pulse height analyser. This analyser works on an anti-coincidence signal from comparing the pulse strength (voltage) with that of two adjustable settings V_1 and V_2. If the signal in the external circuit is greater than V_1 but less than V_2 then it will be accepted. The isolated signal should be proportional to the X-ray flux. The latter can be checked with an instrument configuration (that is insensitive to small focus movements) and varying the tube current or by inserting a series of identical X-ray absorbers placed in the beam. The signal should be proportional to the tube current or to the number of absorbers. For optimum performance this pulse height discrimination should be set very carefully to isolate the signal peak to enhance the dynamic range by removing the noise contribution and maintain proportionality.

Suppose we now have the situation where this proportionality is lost at high X-ray fluxes. This can arise when the electrons and ions cannot reach the electrodes before another ionising photon arrives. The local electric field has therefore not recovered and the gas amplification is less effective and photons are missed. Of course this will happen on a statistical basis at lower levels than for a series of photons arriving at regular intervals. Because of the high fields the pulse width is very narrow in time and the loss in proportionality does not occur until >100,000 photons arrive per second. The loss in proportionality is predictable and can be simply measured by increasing the tube current to give count rates above these values. A proportional detector can be used reliably up to about 750,000 photons arriving per second when this correction is included. This correction is often included in the detector control software. The proportional detector is therefore reliable over ~0.2 to 1,000,000 photons arriving per second, before attenuators are required.

3.4.2. The scintillation detector

The principle of this detector is again reliant on an ionisation process, except this time in a solid. These detectors are composed of a phosphor, usually NaI with ~1% Tl (at least always >0.1% Tl), followed by a photomultiplier tube to obtain reasonable pulse strengths with minimal noise introduction. When an incoming photon ejects an electron to the conduction band of a matrix atom a positively charge hole is formed that drifts to a Tl impurity causing ionisation. Therefore each event creates an electron in the conduction band and an ionised impurity, i.e. Tl^+. The recombination of the electron and the Tl^+ hole will create characteristic fluorescent radiation that has a lower energy than the surrounding matrix. Because the fluorescent radiation has a lower energy than that of the matrix there is very little absorption and can therefore travel large distances, hence the fluorescence decay is determined by the NaI matrix. The whole process takes about 10^{-7} s and the subsequent processes are generally not rate limiting. The possibility of alternative entrapment of the electrons will lead to losses but this is optimised by the correct choice of phosphor and activated impurity.

The phosphor is optically coupled to a photocathode and this in turn is coupled to a photomultiplier. The fluorescent radiation produces electron emission in the photocathode (Cs_3Sb) and the dynodes of the multiplier create electron multiplication to give a measurable pulse that relates to the incoming signal. The variation in the voltages between all dynodes and imperfections in the optical coupling does create some electron loss and light loss respectively and therefore the pulse strength is much more variable than the gas amplification process, consequently it has much poorer energy resolution. The number of fluorescent photons created is related to the incoming photon energy as the electron energy transitions are created as a result of photoelectric absorption and by the Compton process, Chapter 2, figure 2.1. These detectors have very good capture efficiency ~100% of typical X-ray photons, but are bulky in comparison, have a poorer energy resolution and generally higher residual noise levels. At one time the scintillation counters were superior to proportional counters at high count rates and X-ray energies, but with improvement the proportional counter out-performs the scintillation counter in dynamic response due to optimisation of the pulse shape from improved design and improved absorption of incoming photons.

3.4.3. The solid state detector

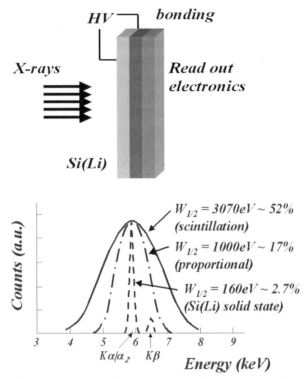

Figure 3.9. (a) The geometry of a solid state detector with a front electrode. (b) A comparison of the energy discrimination for the solid state, proportional and scintillation detectors.

The solid-state detector, figure 3.9a, is less complicated in understanding the basic physics, although the problems in manufacture can be more challenging than those above. An incoming photon produces an electron-hole pair in a semiconductor in the presence of an electric field. The electrons and holes drift to the respective electrodes and are captured. To remove the leakage currents and yet achieve sufficient electron-hole capture speeds is a complex design criterion. The advantages however are that the photon capture efficiency is ~100%, for energies up to ~20keV, and the energy resolution can be very high, ~120eV. To maintain the high field across the semiconductor the resistivity has to be very high and therefore any ionised impurity should be compensated. A common approach is to use Li compensated p-type Si. The Li is a fast interstitial diffuser that neutralises the unsatisfied Si bond associated with the p-type dopant (e.g. B). From figure 2.1 we can see that the major capture process in Si is mainly due to the photoelectric effect and to a far lesser extent the Compton process, i.e. the energy is dissipated by transference to the ejected electron momentum. The amplification of this signal must take place very close to the detector to prevent any disruption of the very weak pulse. The energy resolution for a modern solid state detector is shown in figure 3.9b and is compared with those for a proportional and scintillation detector.

3.4.4. Position sensitive detectors

Position sensitive detectors provide a way of collecting data more rapidly, however in general some of the advantages of counting performance can be lost. If the individual regions are small enough they can show variations in the scattering at the micron level. This latter kind of area detector can be X-ray film (Ilford L4 nuclear emulsion plates) and this has developed grain sizes of less than a micron. The application of fine emulsion film for topography will be discussed through examples in Chapter 4. The pixel size of phosphor and glass capillary optical links to CCDs or MOS type storage arrays is decreasing (presently ~ 5 microns) and can be used for low-resolution topographic imaging. Film has the disadvantages of the development process and CCD and MOS storage systems have the disadvantage of high noise levels and slow read out times (several seconds) for dynamic experiments. The dynamic range of these systems is rather limited $\sim 10^2 / 10^3$ whereas image plates (effectively an electronic version of film) can record intensity ranges over 10^5. An image plate requires reading the X-ray generated colour centres with a laser (promoting relaxation from these metastable states to create measurable light) and is quite a major set-up. However earlier problems with leakage and loss of proportionality with time have been improved but the user must be very careful of artefacts from high-count rate saturation. The pixel size for an image plate is about 200 microns and read out time is a few minutes. This is an area of constant improvements, however some of the more familiar technologies are more suitable for large dynamic ranges. The high noise level in these detectors arises from the lack of discrimination in the counting, leading to an additive noise level that has to be subtracted to achieve the true counts. The intensity measurement is therefore very inferior to the proportional and solid-state detectors.

In order to obtain good position sensitivity the final detection position has to be close to the initial point of ionisation, this favours the proportional and solid-state detectors. These are true photon counters and do not integrate the noise. To obtain the position sensitivity with solid state detectors or phosphor-based detectors the size of the electrode or optical coupling (glass fibre) defines the spatial resolution. The proportional counter based position sensitive detector works on a different principle. Since the gas amplification is very localised yet the signal is received by an extended wire electrode, it is the time for the charge to be swept along the wire or metal electrode that determines the position of the event. The ratio of the electron pulse transit times to both ends of the electrode defines the position for a one-dimensional detector. For a two dimensional detector a mesh of wires running along orthogonal directions and evenly spaced give both co-ordinates.

Clearly the transit time limits the total measurable count rate, since the total time from photon capture to an external measurable pulse is rather long. The total count rate limits can therefore be very low ~20000 photons per second before corrections are required and finally saturation. These flux values are easily achieved with a Bragg peak and therefore considerable care is required if any reliance is needed on the measured intensities. Also the longer the electrode the more severe the problems become. Careful design and attention to perfect uniformity of the wire to maintain good energy resolution are also serious considerations.

For the solid state and phosphor-based detectors the whole basis is quite different and just relates to the scale, since each electrode has its own circuitry. Any detector that becomes too large will have difficulties in that it has to match the geometry of the scattering experiment and even so calibration and checking the response from X-rays with different trajectories should be carried out. Generally they should be curved about the sample position. The attraction of the individual solid state detectors is the small pulse width and fast read out times, thus overcoming many of the disadvantages of the problems of transit time or storage problems. It is important to realise at this stage the usefulness of position sensitive detectors is confined to experiments that require the resolution defined by the detector. The main applications are in scattering from weak single crystals of polycrystalline materials, polymers and samples with limited crystallinity. As the acceptable count rate increases then the application areas will increase. However state-of-the-art solid-state position sensitive detectors have potential to accommodate impressive count rates, comparable to a proportional counter per pixel.

3.5. Incident beam conditioning with passive components

Passive components in this context are defined as those that just restrict the beam spatially and in angular divergence. The wavelength distribution is unchanged unless it has a spatial distribution.

3.5.1. Incident beam slits: Fixed arrangement

Inserting slits into the beam path reduces the divergence of the X-ray source and the beam size arriving at the sample. We shall firstly consider the divergence in the scattering plane, since this has the largest influence on the scattering pattern, then consider the importance of axial divergence out of the scattering plane. We cannot simply state a divergence in the scattering plane without qualifying the shape of the distribution. We will firstly consider the profile of the beam from the simplest configuration of slit and source, figure 3.10a. If we assume that the generation of X-rays is radially uniform so that the X-ray flux reaching the slit is uniform then the profile can be expressed by:

$$I \propto \frac{|\Phi_2 - \Phi_1|}{2\tan^{-1}\left\{\frac{F}{2(L+l)}\right\}} \qquad 3.2$$

where $|\Phi_2 - \Phi_1|$ is the subtended angle at the position y and

$$\Phi_2 = \tan^{-1}\left\{\frac{y+F/2}{L+l}\right\} \text{ or } \tan^{-1}\left\{\frac{y+S/2}{l}\right\} \text{ or } \tan^{-1}\left\{\frac{y-S/2}{l}\right\}$$

depending whether the slit intercepts the view of the focus from the position

y. $\Phi_1 = \tan^{-1}\left\{\dfrac{y - F/2}{L+l}\right\}$ unless the slit intercepts this beam, following the arguments above. Expression 3.2 relates the view of the focus from any position y at a distance l from the slit. The slit and focus are assumed to be co-axial and therefore the normalisation factor gives the intensity that would be registered on this axis at the determined distance from the source.

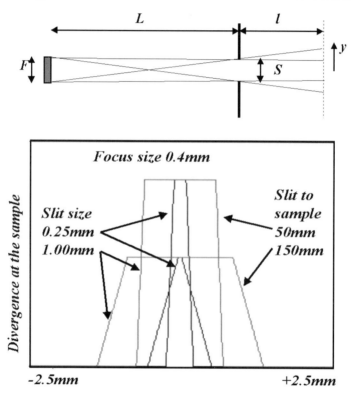

Figure 3.10. (a) The geometry of a single slit and (b) the variation in divergence on the sample for several combinations and a fixed focus size for $L+l = 320mm$.

Figure 3.10b gives the shape of this distribution as we change some of these parameters. It can be seen that a simple divergence angle is insufficient to characterise all these shapes. This shape varies with distance from the slit, therefore the position of the sample with respect to the slit and source will affect the scattered profile. From these profiles we can observe very useful details. Firstly the peak intensity diminishes with distance from the source, because we are capturing a smaller distribution of photon trajectories, this is important for limited sample sizes. The profile closely resembles a triangular, rectangular and more rounded shape as we progress from narrow combinations to larger dimensions. Clearly there is no simple description of the divergence.

Since these profiles are spatial distributions, each point represents a subtended angular divergence of the source and hence the angular divergence varies along y, normal to the axis. This angular divergence will be modified depending on the size of the sample and its projection onto the beam axis normal. It should be apparent that the complexity builds up very rapidly with errors in alignment, variable flux density across the focus and the inclusion of air scattering and absorption. However as the source size is reduced the intensity distribution becomes more rectangular.

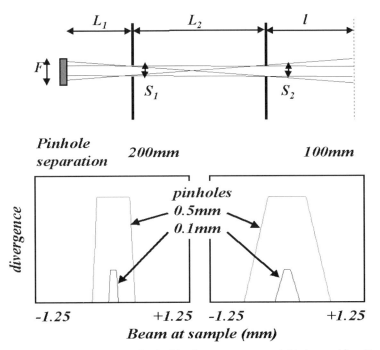

Figure 3.11. (a) The geometry of a double slit arrangement and (b) the resulting divergence on the sample for L_1=70mm and a total focus to sample dimension of 320mm.

124   X-RAY SCATTERING FROM SEMICONDUCTORS

The addition of an extra slit to create a double slit collimator creates an extra degree of freedom and complexity, figure 3.11a. The spatial distribution of intensity arriving at the second slit will be as described in figure 3.11b, however because this has a varying divergence across this distribution the influence of this second slit will in general create a more triangular profile. This triangular profile is indicative of a rapid variation of angular divergence across the emerging beam. If the sample is smaller than the beam size then the divergence will vary with incident angle. If the sample is inhomogeneous then the varying spatial distribution in the divergence could also add complications and it is important to be aware of the variation. The divergence can be made more uniform with a line focus (i.e. a small focus in the scattering plane) however this increases the axial divergence normal to the scattering plane. The axial divergence can be contained as discussed in section 3.5.3.

If the sample is inhomogeneous or we wish to maintain a constant sampled area then we have to control the illuminated area automatically with incident angle. This is discussed in the next section.

3.5.2. Incident beam slits: Variable arrangement

The illuminated area can be maintained constant by varying the width of the slits. This can be done very effectively under computer controlled data collection, figure 3.12a. The mechanical tolerances are very tight and the motion is controlled in general by rotation of the upper and lower parts of the slit. The corrections to the measured intensities to convert them to equivalent fixed divergence and an infinite sample are purely geometrical, depending on the radius of the slits, sample to slit and slit to X-ray source focus distances. The geometry of this arrangement requires a small focus to maintain a controllable divergence and therefore the line-focus option is preferred (figure 3.7.).

As we could see from the explanation of the control of the divergence for a single slit the source to slit and slit to sample distances are all important. High mechanical precision is required for an automatic divergence slit, since the divergence and hence illuminated length for both upper and lower slits are different. The shape of the slits is rather critical to create a good edge to limit scatter from all working openings. Remember of course, the divergence varies over the angular scan range so again we cannot define a simple instrument function.

Suppose that all the X-rays arriving at the sample undergo the same scattering probability, as would be the case for a perfectly random distribution of crystallites (the perfect powder sample). The intensity for an equivalent "fixed" slit arrangement would then relate to the measured profile for this "variable" slit arrangement, with the following expression:

$$I = I_{measured}\left\{\tan^{-1}\left(\frac{\sin\omega}{2\frac{L+l}{x}-\cos\omega}\right) + \tan^{-1}\left(\frac{\sin\omega}{2\frac{L+l}{x}+\cos\omega}\right)\right\}^{-1} \quad 3.3$$

$(L+l)$ is the sample to source distance and x is the length of the sample illuminated. The variation of the intensity is presented in figure 3.12b. It is clear from this variation that the intensity on the sample is heavily reduced close to grazing incidence. At higher incidence angles this intensity increases to a maximum at normal incidence and then declines. The divergence of the incident beam in the scattering plane approximately follows the intensity variation. Therefore any experiment making use of these slits will produce a response that is dependent on the sample, since the divergence is rapidly varying with incident angle, i.e. can the sample make use of this increasing divergence?

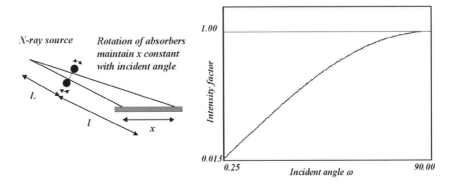

Figure 3.12. (a) The mechanism for obtaining constant area of illumination of X-rays on a sample and (b) the variation of intensity incident on the sample.

From Chapter 2 we could see that the specularly reflected intensity falls very rapidly with increasing angles of incidence and therefore this configuration can be used as a very powerful way of compensating for this effect. However since each beam is specularly reflected we must consider the scattered beam acceptance very carefully. This will be covered in section 3.7.4.1.

3.5.3. Parallel plate collimators

Another method of controlling the divergence of an extended X-ray source is illustrated in figure 3.13a. The slit is composed of a series of thin parallel plates and the divergence is calculated in the same way as for a double pinhole arrangement. The advantage here is that the divergence can be maintained at low levels whilst still maintaining the intensity by using a large X-ray source. The mechanical tolerance cannot be so good as a double slit arrangement and therefore the restriction on the divergence is rather limited. So far we have considered controlling the divergence in the scattering plane, however this arrangement can be used to control the axial divergence (normal to the scattering plane). The parallel plate collimator is rotated through 90° about the X-ray beam axis (Soller slit). This is a very common method used in conventional powder diffraction methods because significant axial divergence introduces a large asymmetric tail on the low scattering angle side of the Bragg peaks, leading to additional complications in fitting the shape and position.

Most applications of this parallel plate collimator considered in this discussion are as an analyser in the scattered beam. This allows the capture of scattering from large regions on the sample and is often used in combination with very low angle of incidence probing beams for analysing thin films. We shall therefore confine the calculation to the accepted divergence from nearly perfect epitaxy and randomly polycrystalline samples. This gives an indication of the resolution and whether any real information can be obtained from the profile shape, figure 3.13b.

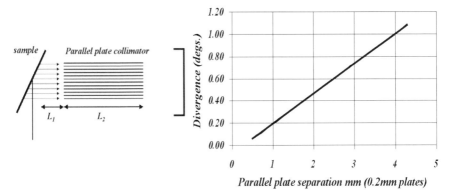

Figure 3.13. (a) The parallel plate collimator used as a scattered beam analyser and (b) how the acceptance varies as function of plate separation. This dependence is virtually linear except at very low separations when the influence of the plate thicknesses becomes significant.

3.5.4. General considerations of slits

The optimum slit configuration to control the axial divergence will depend on the size of the X-ray source and how much intensity is required for the experiment. The double slit combination given in figure 3.11, can be a double pinhole combination (defined equally in all directions) and when used with a point source will produce equal divergence in all directions. When good angular resolution is required in the scattering plane and a line focus is used then the axial divergence is best limited by parallel plate slits figure 3.13, since the projection of the source will be extended in this direction.

Various slit combinations therefore create very different beam profiles and divergences. However the full spectral distribution of the X-ray source is passed and clearly the width of even the strongest characteristic energies can severely broaden the scattered beam and the angle over which the sample will scatter. However we can select a narrow wavelength band by inserting a crystal or periodic multilayer that scatters a specific wavelength band at a specific scattering angle. We shall firstly consider absorbing filters that are frequently used to modify the spectral distribution.

3.6. Incident beam conditioning with active components

An active component is defined as one that changes the spectral distribution. The distribution is changed by removing certain bands of wavelengths by scattering or through the use of absorbing filters. Inevitably the divergence and intensity will also be modified.

3.6.1. Incident beam filters

In Chapter 2 section 2.3 we discussed the absorption of X-rays due to resonance, equation 2.17. What we find is that various elements will absorb X-rays very strongly on the high energy (short wavelength) side of a resonance condition, but above this the absorption is dramatically reduced. By placing an absorbing material of an appropriate thickness, that has an absorption edge very close to the characteristic radiation of the X-ray tube we can change the spectral distribution quite dramatically. The spectral distribution of the X-ray source is given in figure 1.4, where we can see that there are several characteristic peaks due to the Kα doublet and a complex but singular looking Kβ line of significantly lower intensity. Having several characteristic lines adds complications to any diffraction pattern. The energy difference in the doublet is small and high-energy resolution is required to separate these contributions, however we can select an elemental material with an absorption edge just on the high side of the Kβ line. For Cu anode tubes, nickel fits the requirement and for molybdenum tubes zirconium fits the requirement. The spectral distribution is changed so that the Kβ line is almost completely eliminated, but also the broad "white" radiation is reduced and a sharp absorption edge is seen.

For low-resolution measurements of generally weakly scattering material this degree of energy selection may well be adequate, since only the characteristic peaks create measurable scattering. The Kα doublet is then accepted and predictable and for some small crystallites (creating diffraction broadening) or heavily strained material the doublet may not even be resolved.

3.6.2. Incident beam single crystal conditioners

The spectral distribution can be modified in a simple way by a slit and a crystal, figure 3.14. The slit limits the angular divergence, $\Delta\xi$, as described above and therefore from Bragg's equation:

$$\frac{\Delta\lambda}{\lambda} = \cot\theta \Delta\xi \qquad 3.4$$

where $\Delta\xi$ is the angle of divergence defined by the slits and the X-ray source size. The magnitude of the wavelength band-pass, $\Delta\lambda$, is defined by the slit and X-ray source, but the actual centre value of the wavelength, λ, is determined by the chosen Bragg angle, θ. We can obtain the spectral distribution of figure 3.6 with the geometry of figure 3.14 and rotating the crystal. The resolution of the spectral profile is increased as the divergence is reduced. However the divergence cannot be reduced much below significant fractions of a degree and therefore the wavelength band pass is still large and can dominate the scattering widths.

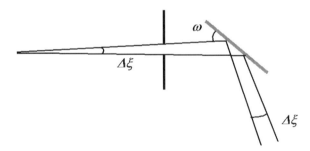

Figure 3.14. The combination of a slit and crystal to limit the wavelength spread.

If however the crystal is kept stationary then a defined wavelength band is selected and each wavelength scatters at an angle defined by Bragg's equation. The double-crystal diffractometer that will be described in section 3.8.1 uses this property. The finite source size creates a blurring of this band-pass because it effectively increases the range of incident angles for each wavelength. However this blurring only occurs over the finite diffraction width and averages out the spatial distribution of wavelengths. The scattered beam from this perfect crystal is hence characterised by spatially distributed beam paths for each wavelength. Each wavelength has a clearly defined direction that is parallel to within the diffraction width of the perfect crystal.

3.6.2.1. Single crystal groove conditioners

From figure 3.15 we can see that the profile of a single reflection from a perfect crystal is sharp and narrow for reflectivities above about 10%. Below this value there are long tails of intensity that would be very evident for experiments using very large intensity ranges. This instrumental artefact introduces considerable aberrations to an experimental profile. These "tails" can be reduced with a grooved crystal, where the X-rays are scattered many times, figure 3.15. From the dynamical scattering model we can calculate the scattering profile from a perfect crystal, figure 3.15 for a single wavelength (each wavelength will have a similar profile, but will be shifted in angular position, ω, for wavelength differences within ~1%). The scattering from the second crystal will have an identical profile for each wavelength.

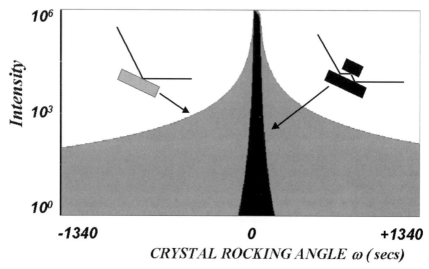

Figure 3.15. The simulation of the scattering from a perfect Ge crystal (220 reflection) for a single reflection and three reflections.

The intensity is a function of the scattering angle and the scattering angle for any part on the curve of figure 3.15 will be the same for each crystal, if the crystallographic planes are parallel to the surface. The resultant profile, figure 3.15, is therefore the product of all the profiles. Since the scattered intensity varies from a value close to the incident intensity to a fraction less than 2% within about 20"arc for the single crystal reflection, the equivalent fraction for N reflections is $(2\%)^N$. This is a useful way of producing very sharp scattering peaks and reducing the instrumental aberrations without significantly reducing the peak intensity. As more reflections take place, the tails are reduced further to values that are completely unobservable. However this is only valid for perfect crystals and no crystals are perfect. Defects resulting from surface damage, point defects and dislocations will create diffuse scattering and broaden the tails of the profiles. The advantages of many reflections can therefore be lost beyond about $N = 3$. Also the maximum reflectivity is slightly less than unity and the overall intensity and shape reduces the intensity quite rapidly especially if the crystals are not absolutely perfect.

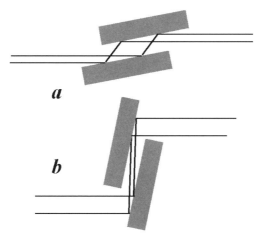

Figure 3.16. Some possible methods of increasing the acceptance and divergence of the beam (a) and decreasing the acceptance and divergence (b) resulting in an increase and decrease in intensity respectively.

The approach above is very useful and can of course be extended to many combinations including crystals cut with non-parallel sided grooves and reflections from planes not parallel to the groove sides. Let us consider the case when the scattering is from planes not parallel to the crystal surface (or groove side). From figure 3.16a we can see that X-rays incident at a grazing angle will arrive at the crystal surface over a large area, resulting in a broad scattered beam after the first reflection. Also the intensity profile over the "rocking angle" is broadened compared with the equivalent reflection from planes parallel to the surface. If the X-rays approach the surface at higher grazing angles and scatter at small angles to the surface then the profiles are narrowed, figure 3.16b. For the condition of low angles of incidence, the angular acceptance of the X-ray beam is enhanced and more intensity is passed. This can be understood from the broadened "rocking angle" profile; although the crystal is not rocked the divergence incident on the crystal is far larger than this broadened "rocking angle." For the geometry of figure 3.16b the intensity passed is reduced. For a grooved crystal with scattering planes inclined to the surface, we again follow this reasoning through the system, although now the options are increased immensely.

To calculate the profile of the intensity through this groove is a little more complex since we have to account for the beam paths rather carefully. For an incident wave-vector arriving at the first crystal at an angle to the surface of ω_I this will create an incident angle at the second crystal of $\cos^{-1}\{\cos\omega_I + 2\sin\theta\sin\varphi\}$, equation 2.85. This in turn will create a scattered wave-vector travelling in a direction parallel to the incident wave-vector on the first crystal. Therefore there is no angular deviation of the X-rays passing through this groove crystal. However the angular spread of the X-rays after the first reflection is reduced and the acceptance of the X-rays on the second crystal matches this acceptance. The beam leaving the second crystal is as divergent as the beam accepted by the first crystal, therefore the intensity is improved at the expense of angular divergence. A factor of 5 in intensity is easily obtainable, compared with an equivalent arrangement when the scattering planes are parallel to the surface. However the crystals have to be larger to capture the same size beam from the X-ray source. A monochromator based on the principle will be discussed in section 3.6.3.

3.6.3. Multiple crystal monochromators

The monochromators described so far rely on the wavelength band being selected by the angular acceptance of the slit. For a typical slit with a divergence of 0.1^0 and the crystal centred on a Bragg peak at $2\theta = 60^0$ with CuKα_1 radiation then the wavelength spread is ~1.55×10^{-4}nm. The broadening of the profile from this wavelength spread varies very strongly depending on the sample scattering geometry. To obtain a versatile diffractometer it is important to have a method of restricting the angular divergences and the wavelength band-pass so the resolution varies only gradually throughout the range where we wish to carry out our experiments. The angular divergence can be kept roughly constant throughout the whole region where we collect data, but the wavelength contribution will always create a varying instrument smearing that increases with scattering angle. This is clear if we rearrange equation 3.4

$$\Delta \xi = \tan \theta \frac{\Delta \lambda}{\lambda} \qquad 3.5$$

Clearly however small the band-pass as θ approaches 90^0 the broadening approaches very large values. For examining high quality crystals and to achieve a versatile instrument it is important to reduce the wavelength band-pass.

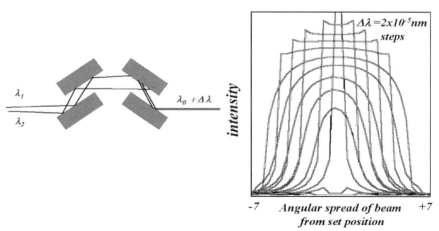

Figure 3.17. (a) The 2-crystal 4-reflection monochromator in symmetrical reflection geometry and (b) the resulting complex divergence variation in arc seconds with wavelength for Ge 220 reflections.

Consider the combination of crystals illustrated in figure 3.17a and the associated X-ray beam paths for a few wavelengths. If we followed Bragg's equation very strictly (i.e. in the kinematical approximation), then only one wavelength can be scattered by both crystals. Because the crystals have finite scattering widths and the beams are divergent then there is a finite wavelength beam-pass. This principle is used in the 2-Crystal 4-Reflection monochromator first proposed by DuMond (1937) and built for a laboratory source diffractometer by Bartels (1983). However the intentions of DuMond to produced narrow profiles was not realised and the widths predicted by Bartels were too excessive. Further calculations by others still over-estimate this value. DuMond's narrowing of the profiles is easily achievable and the monochromator performance can be shown to exceed what Bartels had predicted. The main problem with many calculations is that they are based on convolutions of the angular distributions and wavelength contributions. A more exacting analysis is obtained by considering the angular and spatial distributions of the beam paths.

The approach used here is to calculate the absolute beam direction and position as it passes through the monochromator for each wavelength. Each wavelength from each part of the X-ray source is spatially separated, but the size of the source tends to give an approximately even distribution of wavelengths across the beam in the scattering plane. Let us assume that we have rotated the first U-shaped crystal block to scatter the mid-point of the $CuK\alpha_1$ spectral line, figure 3.17a. The second U-shaped block is then rotated to capture and scatter as much of this distribution of wavelengths and angular divergences as possible. Now any beam of a specific wavelength scattered from the first crystal block will have its own divergence corresponding to the intrinsic scattering profile from reflections in this first block. The direction of this beam will be spread from its central value of $\omega_0(\lambda)$, and we can consider the path of a beam deviated $\Delta\omega(\lambda)$ from this angle. Let us suppose that the second crystal block will accept and scatter X-rays that are incident on it within the intrinsic scattering angle and the mid-range of this corresponds to an angle $\omega_0(\lambda)$. The path of our deviated beam will impinge on the second block at an angle $\omega_0(\lambda)+\Delta\omega(\lambda)$, when ideally it should arrive with an incident angle of $\omega_0(\lambda)-\Delta\omega(\lambda)$. Therefore the optimum scattering angles are moving away from each other at twice the rate of the deviation in angle, $2\Delta\omega(\lambda)$. The width of the intensity profile emerging from the second block will therefore be approximately half that of the intrinsic scattering width. If we now consider a similar analysis for a different wavelength then the second block acceptance angle $\omega_0(\lambda')$ will be offset from $\omega_0(\lambda)$. Wavelengths other than the chosen value will either be deviated rapidly from the scattering condition or go through an optimum position before being filtered out. If we include all these factors then we obtain a family of curves that represent the divergence associated with each wavelength. The resultant distribution of angular divergences for each wavelength centred on the $K\alpha_1$ peak is given in figure 3.17b. It is abundantly clear that we can not separately consider a wavelength distribution and a common angular divergence, but rather each wavelength having its own divergence and intensity distribution. Another important point here is that the alignment of the crystal blocks is quite crucial to maximise on the chosen wavelength and to optimise the angular divergence.

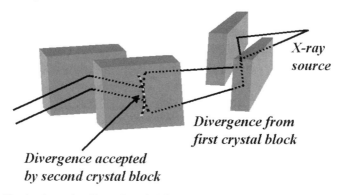

Figure 3.18. A schematic of how the axial divergence is limited on passing through a 2-crystal 4-reflection monochromator.

We have only considered the scattering in the plane of the diffractometer, yet there will be axial divergence that will also influence the profile shape from our experiments. If we consider the path of a single wavelength then it will only be scattered from a crystal surface if it approaches it within the angular acceptance (intrinsic scattering profile width). Therefore if this wavelength is optimised for scattering in the diffractometer plane and the crystal planes are perpendicular to this then those paths out of the plane will be limited to the projection of the intrinsic scattering width. Consider figure 3.18, where the locus of the optimised X-ray beam path is drawn for a single point source at a distance x from the crystal. The beam trajectory is unchanged on passing through the first crystal block in terms of the axial divergence. The projection, θ_I, of the optimum scattering angle, θ_0, for any X-ray beam path will be increased as its trajectory moves out of the plane of the diffractometer. This change in the projected incident angle, $(\theta_I - \theta_0)$, varies as the square of the axial divergence angle out of the plane, χ, equation 3.10. The X-ray beam therefore arrives at the second crystal block with a projected incident angle of $(2\theta_0 - \theta_I)$. Since the projected angle for the optimum scattering of an X-ray beam following this trajectory should be θ_I, the deviation of the actual to optimum incident angle is $(2\theta_0 - 2\theta_I)$. Hence the axial divergence is contained to small values by this monochromator.

$$\Delta\chi = \{[2\theta_0 - 2\theta_1]\cot\theta_0\}^{\frac{1}{2}} \qquad 3.6$$

This is discussed on a purely geometrical argument but for the scope that we wish cover in this book it does explain that the axial divergence is constrained, however for a fuller description we must include all the wavelengths and the alignment of the monochromator, etc.

3.6.4. Multilayer beam conditioners

A laboratory source is divergent and most of the X-rays generated are lost, especially when the divergence acceptance of the monochromator is small or reasonable resolution is required from slits. Ways of capturing this divergence have been discussed for grooved crystals when using crystallographic planes inclined to the surface. However we can make use of a very simple principle of a parabolic reflector. Consider the geometry of figure 3.19. The divergent source is now reflected and concentrated into an almost parallel beam. Of course with X-rays this is not so simple, since they are not reflected efficiently at all angles, section 2.8. This loss of reflectivity is recovered by constructing the parabola as a periodic multilayer whose repeat changes along the surface, Arndt (1990). The satellites from a multilayer composed of strong and weakly scattering materials can give very high reflectivites (~70% or more). To satisfy this condition of maximum conversion of a divergent beam into a nearly parallel beam requires exact manufacture so the period varies to satisfy the Bragg condition for the wavelength of the source. The interface quality between the layers should also be of high quality since any imperfections will reduce the reflectivity.

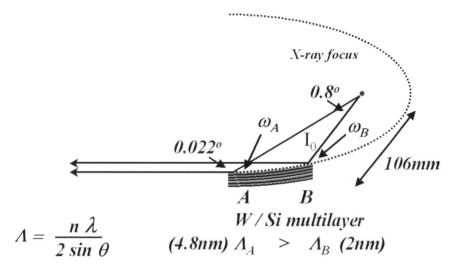

$$\Lambda = \frac{n\lambda}{2 \sin \theta}$$

Figure 3.19. The principle of the multi-layer parabolic X-ray mirror. Λ_A and Λ_B are the periods at the extreme ends of the multi-layer mirror.

To estimate the gain in intensity we can consider the captured divergence, $\Delta\xi$, the degree of parallelism achievable, $\Delta\xi'$ and the reflectivity, R, of the chosen satellite. The gain in X-ray flux (intensity per unit angle) is

$$Gain = \frac{\Delta\xi}{\Delta\xi'} R \qquad 3.7$$

For typical values of $R=0.70$, $\Delta\xi = 0.8^0$ and $\Delta\xi' = 0.02^0$ the flux gain is ~ 28. Clearly as the periodic quality declines, the period variations laterally are not precise and the interface quality is poor the gain is diminished. Also the focus should be ideally point-like and this can be achieved with a line focus with a suitable take-off angle to obtain a 40μm dimension. The X-ray mirror, figure 3.19, can be large in the axial direction (normal to the plane of the figure) to make full use of the line focus. Alignment of the mirror is clearly rather critical but using the prefix arrangements of the diffractometer illustrated in figure 3.1, this is very stable and realignment after removal and return is unnecessary.

Of course the advantages of the mirror lead to disadvantages. We have converted a divergent source into a parallel source and this is not ideal for some applications, especially in diffraction from polycrystalline materials, that may have large scattering widths. The beam has also increased in size and some energy selection will take place. The energy selection does not compare with single crystal reflections at high angles. The mirror utilises a low angle satellite for improved reflectivity and therefore the wavelength band-pass is broad, equation 3.4, and $CuK\alpha_1$ cannot be separated from $CuK\alpha_2$ without great difficulty, but $CuK\beta$ can be virtually eliminated. The axial divergence is increased and therefore the resolution can be made worse, especially in high resolution reciprocal mapping when compared with point focus tube settings. We also have to consider whether the sample can make use of the additional size of the beam to benefit from the increased intensity.

The mirror is therefore another optional component that helps in some analyses, but is not suitable for all. It is though a very useful additional component used in conjunction with slit geometries and multiple crystal monochromators. Increasing the intensity per unit angle has a significant advantage for the latter but now the axial divergence is less controllable and the resolution declines, however for many applications this is a very useful combination for speed. The resolution can be improved with a Soller slit to limit the axial divergence with some loss in intensity.

Toroidal mirrors can also be used to contain the axial divergence (similar to a mirror in scattering and axial planes) but these require a small X-ray source in both planes, i.e. a microfocus source, Arndt, Duncumb, Long, Pina and Inneman (1998). Again we see the intensity can be gained but at the expense of increased beam size. If we use the toroidal mirror to focus onto a small spatial region on the sample then we find the divergence is increased. Clearly there are many possibilities in shaping the beam but it must be adapted to the problem of interest.

3.6.5. Beam pipes

X-rays in general are very difficult to guide unless they are diffracted as above, because the refractive index in all materials is so close to unity.

Limited focusing can be achieved with Fresnel zone plates and these have ring diameters and separations to maintain the phase relationship and acts rather like a lens. However to condense an X-ray beam by guiding has been successful with glass capillaries, Komakhov and Komarov (1990). The procedure relies on the total external reflection of X-rays and gradual curvatures. Consider a bundle of tapered glass capillaries, figure 3.20, where the capillaries are pointing towards the point focus of the X-ray tube. Those X-rays that fall within the angular spread of the capillaries have a good probability of entering them (some will be absorbed by the glass). Since the X-rays will be entering roughly along the capillary axis they have a good chance of travelling through the capillary if they arrive at the walls at a low angle with a high probability of being specularly reflected. Higher energy X-rays have a lower critical angle and therefore there will be some energy filtering in favour of low energy X-rays. In general the capillaries close to the centre of the bundle (where the curvature is less) will transmit all energies and towards the edges the energy distribution will favour the lower energies.

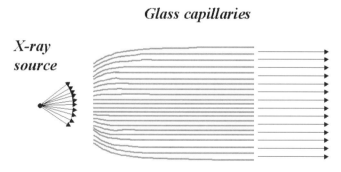

Figure 3.20. The method of recovering the source divergence loss from an X-ray source using glass capillaries. This results in a quasi-parallel beam.

This type of X-ray beam pipe can be used very successfully to capture a divergent source and convert it to a quasi-parallel source. This can prove very useful for low resolution X-ray diffraction studies with large intensity enhancements in terms of flux at large distances from the source. The size of the beam is controlled by the diameter of the capillary bundle, which are generally circular in cross-section giving an even resolution in the scattering plane and axial direction. The divergence is governed by the critical angle

140 X-RAY SCATTERING FROM SEMICONDUCTORS

and will differ slightly for each X-ray wavelength, however as an approximate guide the divergence is twice that of the critical angle.

In some cases a single capillary can be of benefit, either as a straight capillary or one that is tapered to try and induce some spatial focusing. The same principle applies as above, but in this case this beam pipe is useful for transporting intensity with smaller divergence losses than a simple double pinhole for example. The capillaries can be made with diameters less than 100µm.

3.7. Diffractometer options: Combinations with scattered beam analysers

So far we have considered the control of the incident beam and many of these principles can be applied to the analysis of the scattered X-rays from the sample. Again we could consider the analysis in terms of active and passive components, although we will consider combinations that bring both forms together. It is also important to realise that any detector is an active component because it is energy selective. However we will consider in general the description in the type of components as those given in sections 3.5 and 3.6. The important properties that we wish to capture in our detection system are the direction and intensity of the scattered X-rays. If however we are confident about our material and the response is purely related to the incident angle of the X-rays on the sample, as in a rocking curve for example, then a wide open detector set close to where the scattered X-rays are expected may be sufficient. The angle of the incident beam to the surface is the defining parameter. However we must be aware of the response across the detector window and the size of the window and how the detector can modify the measured signal. Except for the latter this will in general be a characteristic of each detector and set-up. This can be easily calibrated, although a well made detector should give a good even response. The scattered beam contains the information that we need to interpret.

To understand how a scattered beam analyser works we have to consider everything that goes before it, the source, incident beam modifications and the sample itself. We now have to consider our sample as a source of X-rays that come from a distributed area and have a range of intensities that vary wildly with incident angle. We also have to be aware of the scattering processes taking place in the sample. Most of the scattered X-rays are coherent and this will tend to dominate the signal observed, however fluorescent radiation, diffuse scattering from sample defects and general scatter from slits and other components need to be understood for a correct analysis. The choice of analysing components will be influenced by these contributions. It is important to realise that as soon as we discuss the scattered beam analyser we are considering the whole X-ray diffractometer and the sample and we cannot consider the analyser system in isolation. We will now cover useful combinations, although many possible combinations exist.

3.7.1. Single slit incident and scattered beam diffractometers

The simplest analyser component is a slit, which in the limit can be the size of the detector window. A slit effectively restricts the angular acceptance of the scattered X-rays and works in conjunction with the illuminated area on the sample. The incident beam conditioning controls the illuminated area on the sample. The sample could be smaller than the incident beam and change the effective beam divergence. The divergence of the beam on the sample and the consequential scattering and analyser acceptance based on the arguments above need to be considered for an exact analysis. Since the complexity of the instrument is rising rapidly, we shall restrict the configurations to common arrangements.

The single slit geometry typical of the Bragg-Brentano configuration, used for studying polycrystalline materials with random orientations, relies on the small projection of a line source to create a divergent beam on the sample. The scattered beam slit (often referred to as the receiving slit) is placed at the same distance as the source from the main diffractometer axis. Because the source is very small compared with the incident beam slit the angular divergence is roughly constant over the sample illumination area and therefore the analysing slit basically defines the resolution. However the analysing slit should not be reduced below twice the X-ray source size for scattering planes parallel to the sample surface. The view of the sample from the slit is total and therefore for the case of a polycrystalline sample with enough crystallites to cover all possible orientations the angular acceptance is a flat response. This configuration is therefore near ideal for most studies on these samples. Deviations from perfectly random distributions can cause many complications, Fewster and Andrew (1999).

Consider a perfect single crystal with the scattering planes of interest parallel to the surface then the scattering will occur at various positions along the sample defined by the Bragg equation. The scattered beam will be defined by an angle of twice the Bragg angle from the point of scatter. Since the point of scatter only corresponds to the diffractometer (goniometer) main axis at one position with a coupled $\omega/2\omega'$ scan (i.e. sample and detector axis rotation maintained in the ratio of 1:2) a signal will only be recorded at this one position. Now if the sample is not so perfect and the structure is mosaic (small crystallites with the crystal planes deviating from the mean position), then the sample can be "rocked." This "rocking" angle that maintains some scattering intensity, whilst the analysing slit is set at the scattering angle for the sample, will be representative of this mosaic orientation spread. This is a useful "rocking" method that gives a quick analysis of epitaxial quality, provided that the mosaic regions are not strained with respect to each other (this would give a different scattering angle for each region). This measure of orientation spread is only relevant to crystallites of reasonable size, i.e. greater than that to create diffraction broadening effects, section 2.7.3. It is also important to remember that the angular spread is a projection of distribution of angles onto a plane normal to the ω axis. Clearly with even the simplest geometry we can start to obtain useful information on "single crystals." Extensions of this geometry, discussed in section 3.7.2, and applications to routine semiconductor analysis will be given in Chapter 4.

The single slit Bragg-Brentano geometry described above is only optimised for the sample surface and the scattering planes being parallel and the incident angle ω being half the scattering angle, $2\omega'$ as measured by the diffractometer. As we move away from this condition the projected view of the sample as seen by the incident beam slit and the scattered beam slit differ and the angular acceptance no longer matches the angular divergence. The angle through which the analyser assembly can be moved without loss of intensity is greater and therefore the resolution declines. However this understanding is only relevant to a perfectly randomly orientated set of crystallites. For a perfect single crystal the scattering is located in a region comparable to the focus size. This is the only region that is satisfied by the Bragg condition and therefore the scattered beam direction is well defined, since it comes from a small region on the sample, consequently the resolution is not significantly impaired. Of course this argument is for a monochromatic source and in reality the spectral distribution creates a large region on the sample from which scattering is possible and this defines the resolution limit in this case.

3.7.1.1 Applications in reflectometry

Provided that the instrument is well aligned this combination can produce a medium resolution diffractometer suitable for reflectometry measurements for example. Suppose that the incident beam can be narrowed down to give a divergence of $0.03125°$ ($1/32°$) we can then derive the optimum receiving slits for the various geometries. For the specular scan $2\omega = 2\omega'$ precisely the angular acceptance varies very slowly with angle and all the intensity is collected with a 0.7 mm slit at the same distance, 320mm, as the source from the sample centre. As we scan away from the condition for which $2\omega = 2\omega'$ then the resolution is changed. For $2\omega > 2\omega'$ the divergence slit is too large and excess residual scattering can reach the detector and for $2\omega < 2\omega'$ this analysing slit does not pass all of the expected scattered beam because the illuminated area has enlarged. The acceptance slit can be narrowed below the optimum without significantly changing the profile shape, except to reduce the diffuse scattering, but it will reduce the signal strength and modify the dynamic range to a small extent because of the varying instrument probe at small angles. To expand on the latter comment we have to recognise with this geometry that the scattering angle is related to the position on the sample, where the scattering occurs. A very large sample for example may create a similar angular distribution of specular beams as a small sample but the beams are very well separated spatially and this defines the requirements for the slit. An alternative approach is to reduce the incident beam slit to a very small value and place a double slit at the acceptance end or just use a very small sample! These approaches are methods for defining the area on the sample and there have been many ways to achieve this. Perhaps the simplest way of defining the beam on the sample is a knife-edge placed close to the sample. The illuminated area of the sample that can be seen by the scattered beam is then defined geometrically, figure 3.21. The intensity modification from the use of a knife-edge is approximately given by

$$I = d\cos\omega(2 - \left(\frac{d}{L}\right)\sin\omega) \qquad 3.8$$

where d is the distance of the knife-edge to the surface and L the distance to the source. Clearly if the knife-edge is close to the sample the intensity falls and the intensity is roughly constant over small values of ω. The intensity falls by 0.5% out to $\omega = 5^0$ and is independent of d (apart from the overall intensity) unless the beam is allowed to pass over the end of the sample. To prevent this the edge should be brought closer to the surface.

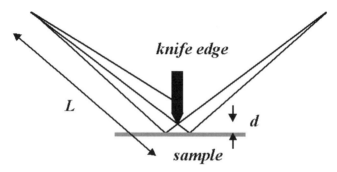

Figure 3.21. The principle behind defining the region on a sample with a knife-edge.

All these methods have their merit and depend on the nature of the problem, the precision required and the quality of the sample.

To set up the diffractometer for undertaking a reflectometry (specular) scan, the sample must first be centred in the X-ray beam. This is achieved by translating the sample normal to its surface to cut the narrow beam in half. To ensure the surface is parallel to the incident beam it should be "rocked" in ω to maximise the intensity, it is this maximum intensity that should be halved. If the profile is rather broad the sample can be moved to reduce the intensity to 10% to do this; then driven back to achieve the 50% condition. Since this is a specular scan the incident angle must equal exactly half the scattered beam angle and this is achieved by setting the detector to an angle between 1° to 1.5°. If the sample is now "rocked" about an angle half that of the scattering angle then as it scans through the specular condition a strong signal should be observed. This is the position to set the incident beam angle, ω. The two axes should now be coupled by defining the ω angle to this new value. The specular scan is obtained by scanning both axes in the exact ratio of 1:2.

3.7.2. Enhancements to the single slit incident and scattered beam diffractometers

Two problems with the geometry described above come from unwanted scatter. The angular acceptance of the analysing optics can be improved with the addition of a second set of slits (anti-scatter slits). These can be set to only allow scattering from the direction defined by the illuminated region of the sample and the receiving slit. This improves the signal to noise ratio. These slits can be automatically controlled for maintaining the Bragg-Brentano geometry for perfectly randomly orientated polycrystalline materials.

The other source of unwanted scatter is associated with the sample from fluorescent X-rays created by the incident beam. This is clearly a sample characteristic, some of which can be removed by the detector pulse height analysis, however fluorescent X-rays close in energy to the coherent scattered beam will need to be removed. For example, samples containing Fe studied with CuKα radiation will create a very large dominating background intensity. The only way to reduce this is to increase the energy resolution of the detection system and a very simple way to achieve this is with an analysing crystal. For the study of polycrystalline materials graphite with a controlled mosaic spread works very well, although a factor of 3 in intensity can be lost. This graphite diffracted beam monochromator also removes the Kβ radiation. A typical graphite crystal will have a mosaic spread of ~0.5°. Ideally the monochromator crystal should fit on a radius including the analysing slit and detector window, however a flat crystal is acceptable because of the mosaic spread in these crystals.

A typical Bragg-Brentano diffractometer will contain all the components described and this can be a useful instrument for preliminary studies on semiconductors including, approximate alloy composition measurement, superlattice analysis, crystal quality analysis and low resolution reciprocal space mapping. Examples of these will be given in Chapter 4.

Further resolution improvements can be made with perfect single crystal monochromators in the incident beam. These crystals are usually Ge and are bent to focus the spectrally pure wavelength ($K\alpha_1$ and no $K\alpha_2$ component) onto a slit that effectively becomes the source in the Bragg-Brentano geometry. The alignment of this arrangement is very critical and quality design is required to maintain the stability for everyday use. This can prove to be a very useful source for low-resolution analysis of nearly perfect crystals.

3.7.3. Double slit incident and parallel plate collimator scattered beam diffractometers

The discussion above touched on a few problems associated with single slits and controlling the view of the illuminated area from the incident beam and the scattered beam slits. This is a limitation for studying scattering planes inclined to the sample surface from anything other than highly perfect crystals. This can be partially overcome with the geometry illustrated in figure 3.22. The parallel plate collimator acts as a double slit assembly but extends over an area of the sample that can be very large. Hence for low angles of incidence the whole scattered beam can be captured, for example. The resolution of the incident beam optics is really quite easy to modify, however the manufacturing tolerance (width of blades and separations) make for less easily controlled scattered beam optics. The resolution of this combination is therefore rather poor but is much more even than the single slit geometry described above from scattering planes inclined to the surface.

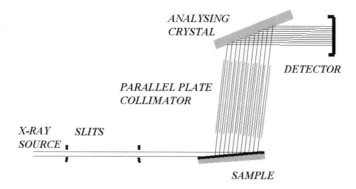

Figure 3.22. The use of a parallel plate collimator for capturing the scattering from large regions on a sample and still maintaining a reasonable resolution.

This geometry does give enhanced signals from very thin polycrystalline layers by using incident angles of ~1.5° and just scanning the detector axis, 2ω'. The advantages of low incidence angles is that the substrate contribution to the scattering is often reduced, although the crystallites could well have preferred orientation that could alter the intensity distribution in the scattering. This geometry is generally applicable to finding the positions of peaks and not the detailed shape. The alignment requirement for this combination is not very critical since the resolution is low and the peak positions are defined by the direction of the scattered beam, rather than the relative positions of the sample and analysing slits.

3.7.3.1. Enhanced double-slit incident and parallel-plate collimator scattered beam diffractometers

As with the arguments of unwanted scattering described above, the addition of an analysing crystal after the parallel-plate collimator can improve the signal to noise ratio and remove unwanted characteristic wavelengths, e.g. CuKβ. However because the better angular acceptance of this collimator compared with a single slit the improvement is less marked. The graphite or lithium fluoride crystal placed after the parallel plate collimator can define the resolution unless the divergence is smaller than the angular spread of the crystal mosaic blocks. If that is the case then the advantages depend on the application very strongly.

3.7.3.2. Applications for low-resolution in-plane scattering

Although in general the geometry above is used in combination with a line focus and slits (i.e. considerable axial divergence is acceptable) there are advantages in using a double pinhole collimator or beam pipe capillary for some applications, section 2.9. The configuration is identical to that of figure 3.22 except that the double pinhole replaces the double slit assembly and the X-ray focus is rotated to give a point focus. The control over the axial divergence is imperative for examining scattering planes normal to the surface plane, figure 3.23. The procedure for setting up such an experiment is as follows.

The choice of scattering plane, approximately normal to the surface should be decided first, so that the scattering plane normal approximately bisects the angle between the incident beam and the expected scattering beam direction. We will see later that a large tolerance is acceptable. The sample with its surface parallel to the normal scattering plane is moved to cut the incident beam in half. Remember to rock the sample about, ω_{90}, an axis normal to the incident beam direction and parallel to the surface to ensure the sample surface is parallel to the incident beam direction, this occurs when the maximum intensity is measured at the detector. The large detector window can accommodate scattering from the sample up to several degrees out of the normal plane of scattering. A beam-stop slit should now be placed in front of the detector to just isolate and remove the incident beam passing over the sample, figure 3.23. The sample can now be scanned about the axis ω_{90}, and a reflectometry curve obtained. The resolution of this depends on the collimation, but the angle of inclination of the beam to the surface gives an indication of the depth from which the scattering is obtained, figure 2.26. The finite divergence of this incident beam gives a distribution of depths from which the scattering originates. However these angles are always small and generally just above the critical angle for the sample where there is measurable specular scattering. We now know that the X-ray beam is penetrating into the sample and we can find a reflection in the normal way with an open detector (or parallel plate analysing slits if we are confident of the scattered beam direction).

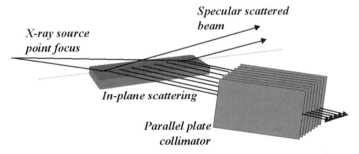

Figure 3.23. The application of the parallel plate collimator for analysing scattering from planes normal to the surface plane.

Suppose now we have found a reflection then it will be clear that the incident beam will be distributed over a large length of the sample, because of the low angle of incidence. The analyser acceptance should therefore be small in angle for resolution but accept the scattering from a large area. This configuration is ideal for the parallel plate collimator. The alignment of the scattering planes parallel to the main diffractometer axis is less clear since the X-rays come in at an angle inclined to the surface and out at an angle inclined to the surface, where the relationship between the former and the latter is given by equation 2.144. Hence the scattering is inclined to the normal scattering plane. If the incident angle is kept small with respect to the surface then we should only be concerned with the small deviation in the projected angle due to the scattering out of the plane. This can be equated to the tilt error, $\Delta\chi$ in sections 3.8.1.2 and 3.8.3.2 where in this case

$$\Delta\chi = (2\omega'_{90} - \omega_{90}) + \tan^{-1}\{\cos 2\omega' \tan \omega_{90}\} \qquad 3.9$$

where the various symbols follow that of section 2.9.

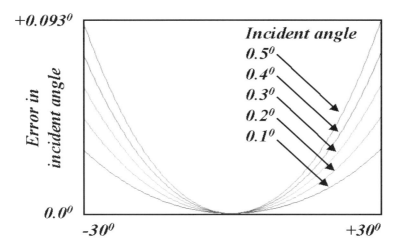

Figure 3.24. The associated error in the incident angle with respect to the surface when the sample is not rotated about the surface normal for in-plane scattering. The deviation in the angle axis of rotation from the ideal is simply the incident angle.

Suppose now that we wish to scan over a very large angle and the rotation about the surface normal does not coincide with that of the main ω axis. The error that should concern us is the change in incidence angle since this leads to a change in penetration depth. This has been calculated for a large deviation from the optimum position and several initial incidence angles and the extent of the error introduced are shown in figure 3.24. It is clear that for small incidence angles large angular ranges are perfectly acceptable and as the initial incident angle is increased the error is still quite acceptable when the beam divergence and uncertainties about sample flatness are considered. Therefore creating an extra axis to precisely rotate about the surface normal is unnecessary for this type of application.

This geometry can easily be used for examining peak positions for determining approximate in-plane lattice parameters or relative orientations between layers and substrates. An example of this application is given in Chapter 4.

3.7.4. Diffractometers using variable slit combinations

As described in section 3.5.2 the variable divergence slit maintains a constant area of illumination on the sample. Also as with any single slit and small source size combination the divergence is constant over the area of illumination. However the angle that each path makes with the sample surface varies. For a randomly orientated polycrystalline sample at any one angle (when $\omega = (2\omega')/2$) it satisfies the Bragg-Brentano geometry, i.e. scattering from atomic planes with the same spacing that satisfy the orientation conditions will come to a focus at a point at the same distance as the focus from the source. Therefore for perfectly orientated polycrystalline samples the source size, the analysing (or receiving) slit and the axial divergence determine the resolution. The axial divergence can be controlled by parallel plate Soller slits, section 3.5.3. Since the X-ray flux reaching the sample varies with incident angle we have to apply the correction given in figure 3.12b, provided that the analysing slit accepts all the scattered X-rays. As with the single fixed slit arrangements, section 3.7.1, this geometry can be used for near perfect samples with little loss in resolution with $2\omega \neq 2\omega'$.

3.7.4.1. Applications in reflectometry

Clearly the idea of an increasing flux with incident angle is attractive for reflectometry measurements but again we must be aware of the changing acceptance required for the analysing slits. If the analysing slits are too wide to accommodate scattering at high angles then the background scatter rises. If a double slit combination is used for the analysing arrangement then it is important to account for the varying flux viewed by this combination and the acceptable divergence. This is where the single slit is a problem despite accepting the same divergence across the whole surface of the illuminated region, since this only works if each region scatters at the scattering angle. The specular reflection will scatter at $2\omega' = 2\omega$ regardless, i.e. it is defined by the incident angle. We can solve this by using a parallel plate collimator with a chosen acceptance angle that does not fall below that of the incident beam divergence at high angles.

3.8. Scattered beam analysers with active components

Although we have considered analysing crystals in conjunction with slit based systems in the last section the primary component defining the resolution function of reciprocal space probe is the slit. In this section the primary components defining the resolution are crystals. The first to be discussed is the simplest and has been a very useful diffractometer in the analysis of semiconductors.

3.8.1. The double crystal diffractometer

The simplest high-resolution diffractometer is the double crystal diffractometer and some of the concepts were discussed in section 3.6.2. Consider the case when the sample scatters at a similar Bragg angle to the incident beam conditioner then all the wavelengths will be scattered at the same angular rotation with the geometry given in figure 3.25. As the sample is rotated further from the matching Bragg angle condition, the wavelength contributes more and more to the broadening. This geometry is the principle of the double-crystal diffractometer (Compton 1917). This geometrical arrangement is really limited to near perfect crystals that are not bent and match the crystal incident beam conditioner. For routine analysis of near perfect semiconductors a double-crystal diffractometer can collect data in a few minutes or even seconds. The intensity can be very high and therefore the possibilities of significant scatter reaching the detector can lessen the dynamic range. The unwanted background scatter can be dramatically reduced with a slit between the two crystals or a knife edge just above the surface of the region where the incident beam arrives at the sample, figure 3.25. An example of this type of analysis will be given in Chapter 4.

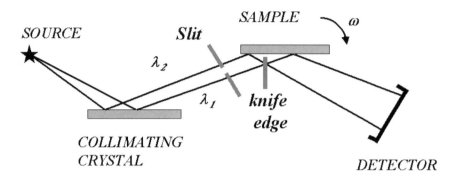

Figure 3.25. The geometry of the double crystal diffractometer with various modifications to improve the performance.

The contributors to the broadening of the profile are the intrinsic scattering width of the first and sample crystals, the wavelength dispersion from imperfect matching of the Bragg angles, crystal curvature and imperfect material. Restricting the beam width on the sample can reduce the influence of curvature. Because of the simplicity of the instrument replacing or realigning the first crystal to match the scattering angles of the collimating crystal and sample can be fairly rapid, Fewster (1985). To overcome the frequent exchanging of the first crystal the 2-crystal 4-reflection monochromator can be used, section 3.6.3. The principle of data collection is similar to the double crystal diffractometer. The alignment procedure for the sample crystal will be discussed and is applicable to all high-resolution diffractometer analyses.

3.8.1.1. Alignment of high resolution diffractometers

The critical consideration is to ensure that the scattering plane normal is perpendicular to the sample rotation axis, ω and in the same plane as the incident beam. The scattered beam should then occur in the plane normal to the sample rotation axis (the diffractometer plane) and the angles measured should be true angles and not projected angles, figure 3.26. If this condition is not satisfied then the measured projected angle will be different from the true angle but also the way in which the reciprocal lattice point interacts with the instrument probe will broaden the profile. For even moderately precise measurements some alignment is necessary, for quality measurements it is imperative.

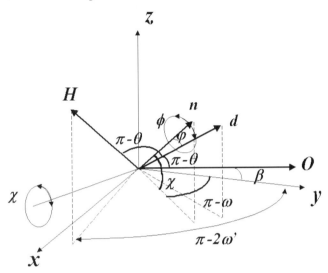

Figure 3.26. The angles to be considered for deriving the errors associated with sample and diffractometer misalignments.

In the case of most high-resolution diffractometers we have two methods for bringing the scattering plane normal into the diffractometer plane depending on the scattering plane of interest. If the scattering plane is approximately perpendicular to the ϕ axis, then rotating ϕ has little effect but rotating χ is very effective. Generally the latter axis is the refining axis and the former will give an approximate setting for planes inclined to the surface. As a general rule for detailed analyses it is worth setting the azimuthal direction with a ϕ rotation to align another reflection that defines this direction (e.g. using the 444 reflection of a (001) orientated sample to set the <110> along the incident beam direction for $\omega = 0$). From figure 3.26 we can determine the error in the ω angle for various errors in the tilt angle, χ.

$$\Delta\omega \sim \frac{\chi^2}{2}\tan\theta\cos^2\zeta \qquad 3.10$$

where ζ is the angle the scattering plane makes with the ϕ axis in the plane of the diffractometer. This varies as the angular separation increases. Hence for an error of 1^0 in the tilt angle the angular mismatch is in error by 1"arc in 300" for the 004 reflection from GaAs, Fewster (1985).

3.8.1.2. Applications of the double crystal diffractometer

Since the instrument has two axes it is sometimes referred to as a double axis diffractometer, although during data collection only the sample rocking axis, ω, is rotated. The wide-open detector is placed to receive the scattered X-rays from the sample as it is rocked, sweeping through the diffraction condition. The scattering profile from a perfect crystal can be simulated relatively easily and to a good approximation the instrument function is defined by the profile of the collimating crystal; correlation of these two profiles should match the measured profile. The collimating crystal is stationary and each wavelength will be collimated to within its own intrinsic scattering profile (the beam on the collimating crystal must be from a finite size source, so effectively each illuminated part of the crystal receives a convergent beam). The angle of convergence must be larger than the intrinsic scattering width so that it is equivalent to rocking the collimator crystal at each angular movement of the sample. This mimics the process of correlation.

The double crystal diffractometer is limited to fairly perfect crystals and this can be a serious limitation if the sample is bent (creating a range of allowable incident angles) and there is no discrimination of the scattered beam. To overcome some of these problems the triple crystal diffractometer was developed.

3.8.2. The triple crystal diffractometer

The triple crystal or triple axis diffractometer is an extension of the double crystal diffractometer and can be considered using the simple ray-tracing argument of the double crystal diffractometer. The analyser crystal matches the collimating first crystal and will now only pass those scattered X-rays from the sample satisfying the Bragg condition of the analyser. Since all the axes are independent scanning through the Bragg condition from the sample is very complicated, the sample and analyser crystal need to be rotated, Iida and Kohra (1979). This instrument led very quickly to undertaking reciprocal space maps by setting the sample rotation and scanning the analyser crystal axis, resetting the sample and scanning again, etc., until a full two dimensional distribution of intensity is recorded. With computer control both axes could be moved simultaneously. The complexity of the instrument movements and the lack of versatility, it was very complicated to change to different scattering planes (the whole instrument needed to be rebuilt), meant that it only existed in a few laboratories around the world, while the double crystal diffractometer flourished.

3.8.2.1. Applications of the triple axis diffractometer

The instrument can be used as a precision lattice parameter comparator. This requires specialised techniques that have been reviewed recently by Fewster (1999) and previously by Hart (1981). The problems of resetting the axes and realigning the whole instrument for each region of reciprocal space were overcome with multiple-crystal diffractometry.

3.8.3. The multiple crystal diffractometer

The versatile monochromator described in section 3.6.3 largely overcomes the problems of the double and triple crystal diffractometer as regards to changing the first crystal, however a sample that is bent or imperfect will still give diffraction profiles that are difficult to interpret. Basically we wish to create an instrument with an incident beam well-defined in direction and wavelength spread and an analysing system that is also well defined so that no realignment is necessary. This would give the user freedom to analyse any set of scattering planes without changing the instrument. The instrument that satisfies these criteria is given in figure 3.27, Fewster (1989). The monochromator controls the scattering plane divergence and the wavelength dispersion giving a well-defined incident beam. The analyser crystal is placed on an axis common with the sample rotation and therefore always points at the sample. The analyser crystal only passed scattered X-rays that are coming from the sample in the specific direction defined by its rotation about the common axis.

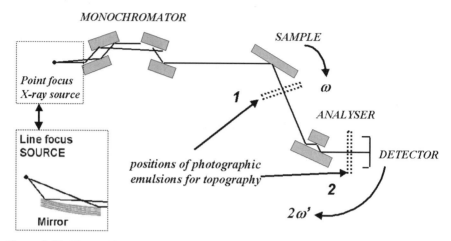

Figure 3.27. The multiple-crystal diffractometer indicating the optional positions for topographic emulsion film and the exchangeable X-ray mirror and point focus arrangements.

The monochromator has been described in detail in section 3.6.3 and it was clear that the wavelength band-pass and divergence is a very complicated function. The analyser crystal is far less complicated but is limited to three internal reflections for several good reasons. Firstly from the earlier arguments extending the number of reflections creates little return because of the imperfections present in the most perfect crystals. Secondly the advantage of offsetting the detector (this occurs with odd numbers of reflections) reduces the chance of the directly scattered beam from reaching the detector. This gives a very good compromise to maintain high intensity and low residual background. From calculations of ideal crystals the contribution from the analyser and the monochromator is exceedingly small and unobservable over a dynamic range of 10^9. In practice the practical limit can be checked by scanning the analyser and detector arm through the incident beam direction, figure 3.28. Because of the very high intensities this must be done with absorbers in the beam path.

From the arguments so far we should consider the angular acceptance of the analyser in conjunction with the divergence and wavelength dispersion of the monochromator. For the sake of a conceptual understanding we shall consider the divergence passing through the monochromator to be half that of the intrinsic full-width at half-maximum intensity of the first reflection in the monochromator. If we now have a sample that has a delta function response then clearly rocking the crystal will give a spread in scattered beam directions with an angle equal to that of the incident beam divergence. Therefore our analyser would appear to need to have an acceptance half that of the intrinsic scattering width of the first reflection from the monochromator. However at this point we cannot take this approach any further because the complexity of all the various interactions do not easily lend themselves to descriptions in "real space." We will refer to figure 3.5 to illustrate the basics of the instrument function.

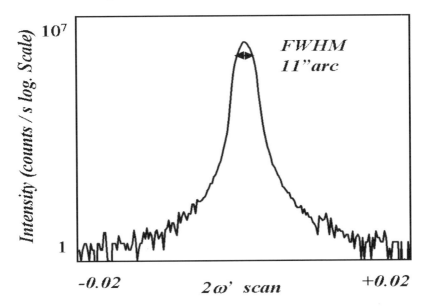

Figure 3.28. The variation in intensity on scanning the analyser and detector through the incident beam direction using the point focus geometry of figure 3.27. Air scatter heavily influences the background close to the position.

The instrument function or probe is the overlap of the angular divergence of the monochromator, the wavelength dispersion and the angular acceptance of the analyser. We will neglect the axial divergence for this argument, since the basics were discussed earlier and the analyser does little additional discrimination of the axial divergence. From our discussion on the theory, Chapter 2, it is clear that scattering from a sample

158 X-RAY SCATTERING FROM SEMICONDUCTORS

with a flat surface will create a profile that can be fully captured by scanning our probe normal to the surface, along s_z. If our sample is perfect then this profile is a line of no width within the assumptions of the dynamical theory. However to capture this information we must have an X-ray source that can only exist with a finite wavelength distribution and divergence to satisfy our understanding of quantum physics. Therefore if we superimpose our probe on this profile we can see that the smearing effect from collecting the data is strongly influenced by the angle of the smearing of the various components, wavelength, divergence and acceptance. The angular spread of these components is a function of the angles ω, $2\omega'$ and λ, figure 3.29. We can of course choose appropriate reflections and minimise these smearing effects.

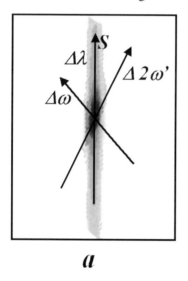
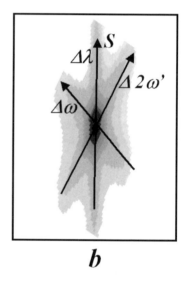

a *b*

Figure 3.29. The simulated reciprocal space map for a perfect Si sample (004 reflection) with the multiple-crystal diffractometer (a) and a triple-crystal diffractometer (b). The surface truncation rod, or dynamical streak is along the scattering vector direction because the simulation is for scattering planes parallel to the surface. If the scattering is from inclined planes then the dynamical streak is inclined to this scattering vector by the inclination angle.

Consider an imperfect sample that gives a distribution of scattering in reciprocal space as in figure 3.30a: if it was perfect it would have a very small width in ω, or normal to the dynamical streak, figure 3.29a. Now the probe is the region defined by the spread in the dimension $\Delta\omega$ and the angular acceptance of the detector $\Delta 2\omega'$, figure 3.5. If the detector is wide open then the whole line represented by $2\omega'$ will represent the detector acceptance. The data collection with this geometry is carried out by "rocking" the reciprocal lattice points in ω through the probe to obtain a "rocking curve."

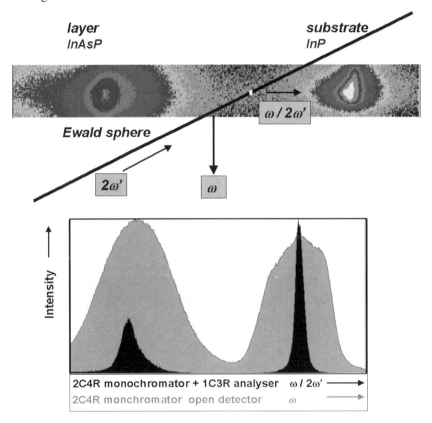

Figure 3.30. (a) The interaction of a section of the Ewald sphere with the reciprocal lattice points from a semiconductor structure, indicating the size of the scattered beam acceptance for the open detector and multiple-crystal geometry. (b) The difference between the open detector scan and that with an analyser.

It is now possible to see the difference in the "rocking curve" with an open detector for an imperfect sample (scanning in ω) compared with a scan combining $2\omega'$ and ω along s_z with the multiple crystal probe (illustrated in figure 3.30a). The influence of bend and other imperfections can be made

160 X-RAY SCATTERING FROM SEMICONDUCTORS

orthogonal to this scan and so just the strain is modelled for example, figure 3.30b. Alternatively the whole map of figure 3.30a can be projected onto the s_z and modelled, thus removing all the influence of tilt and bend but capturing all the intensity from the probed region of the sample.

Let us consider the details of the probe for mapping the intensity in reciprocal space. If the sample is perfect then the addition of an analyser has little or no effect, since the probe may be larger than the intersection of the monochromator and the sample profile. Hence we can see that when the instrument function is too large it is just not used to the full and we have to be aware of this in our modelling. For perfect samples, therefore the analyser can have a large acceptance and a simple rocking curve similar to the data collection method of the double-crystal diffractometer will collect all the data we need to model the data and extract information. If the detector window is smaller than the 2ω' range of the scattered beams, then we can also scan the detector, to ensure all the intensity is captured. This two axis scan also has the advantage of ensuring that the same region of the detector collects the data at each point for reflections with planes parallel to the surface. For large rocking curve scans scanning the detector (2ω') and rocking angle (ω) along s_z, figure 3.3 will also help to solve the problem of the finite size of a detector window.

3.8.3.1. General considerations of data collection

The real benefit of the multiple crystal diffractometer is in the study of imperfect samples, i.e. real samples. Imperfect materials will have regions that can cause diffraction broadening, they may be orientated with respect to each other, the sample could be bent, there could be many defects that create diffuse scattering, etc. The importance of the probe size for these studies now becomes rather crucial. The profile of imperfect materials will be broadened considerably and the range over which the incident beam will create scattering will be larger than the probe dimensions, figure 3.30a. Clearly if the analyser acceptance is large then the intensity assigned to the position (ω, 2ω'), the centre of the probe, will be the sum of all the contributions. These contributions can include bend, mosaic spread, finite size effects and strain with little chance of isolating these effects. The smearing effects of that arrangement will therefore confuse the contributions of the various sample properties that we wish to determine.

The ideal combination could be for the analyser acceptance to match that of the incident beam divergence. As we have seen when we discussed the monochromator performance the divergence is a very complex relationship of wavelength, the various contributions depend on the angles of ω and 2ω' as well as their profile and widths. Also from our theoretical analysis we know that the strongly scattering planes have broad profiles. So we have several possibilities and these depend on the material to be analysed and the type of experiments. We can have an acceptance twice that of the divergence to maintain good intensity, increase the divergence with asymmetrically cut first crystals or reduce the acceptance with asymmetrically cut crystals. Without modelling the whole system including a typical sample estimating the most beneficial combination is only guesswork.

3.8.3.2. Alignment of multiple crystal diffractometers

As in any high precision measurement the alignment of the scattering planes to coincide with the diffractometer plane is crucial to prevent the measurement of projected angles. This was partly covered in section 3.8.1.2, but with the addition of an analyser we do have more possibilities. Generally the alignment and setting of the sample on the diffractometer should not be too different from that suggested for the double-crystal diffractometer, however this only brings the scattering plane normal into the plane of the diffractometer (preferably by the minimum ω method) and optimises the ω rotation. This is obtained with an open detector since the scattering does alter the scattering angle. When the analyser crystal is substituted for the open detector then we know the sample is scattering and therefore we need to scan the detector / analyser axis (2ω'). The maximum intensity gives us the setting for this 2ω' axis and then we are in a position to undertake a scan normal to the surface plane (along s_z, figure 3.3) or obtain a reciprocal space map. A diffraction space map is obtained by scanning ω and 2ω' in a 1:2 step ratio and offsetting ω by $\delta\omega$ before the following scan. This gives a radial sector of reciprocal space, figure 3.31a, and can then be converted in software to form a reciprocal space map. Alternative maps can also be collected along any direction by using something other than the 1:2 step ratio and offsetting in $\delta\omega$ and $\delta(2\omega')$ dependent on the particular region of reciprocal space of interest. These

162 X-RAY SCATTERING FROM SEMICONDUCTORS

maps can then be converted to reciprocal space using the relationships given in equation 3.1. The data can also be collected directly in reciprocal space (e.g. along s_z and s_x) and this can have advantages in interpretation; compare with figures 3.4a, b and c. The angles that the diffractometer needs to be moved to, in terms of the reciprocal space co-ordinates, are given by:

$$2\omega = 2\sin^{-1}\left(\frac{\lambda\sqrt{\{s_z^2 + s_x^2\}}}{2}\right)$$

$$\omega = \tan^{-1}\left\{\frac{s_x}{s_z}\right\} + \omega'$$

3.11

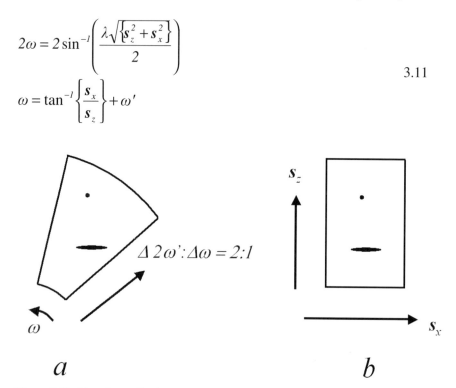

Figure 3.31. Two data collection strategies for reciprocal space maps; (a) collecting a segment of reciprocal space by defining angular movements and (b) collecting the data directly in reciprocal space.

The reciprocal space map can therefore be obtained in any parallelogram shape in reciprocal space, figure 3.31b. We have taken the reciprocal space co-ordinates defined related to the surface and this can be useful for wafers, however this may not be very convenient for samples with irregular surfaces or structures that do not have a simple direction parallel to the surface. In these cases it may be more convenient to define s_z parallel to a crystallographic direction, in which case a reflection along the defined s_z should be used to define the ω direction such that $\omega = 0.5(2\omega')$, and similarly the $\chi = 0$ should be set at this optimum scattering condition. The translation between the diffractometer and reciprocal space coordinates can be obtained by creating an orientation matrix, Hamilton (1974).

To put reliance on the measurement of the various angles we need to determine the dependence of these on the various diffractometer angles. The details of the geometry will not be repeated here but the errors in the measured values of ω and $2\omega'$ due to sample misalignments and the incident beam not being parallel to the diffractometer plane are given by

$$\omega = \sin^{-1}\{\frac{\sin\theta - \cos\Delta\chi \sin(\chi - \Delta\chi)\sin\beta}{\cos(\sin^{-1}(\cos\Delta\chi \sin(\chi - \Delta\chi))\cos\beta}\}$$

$$-\cos^{-1}\{\frac{\cos^2\Delta\chi\cos^2(\chi-\Delta\chi) + \cos(\sin^{-1}[\cos\Delta\chi\sin(\chi-\Delta\chi)]) - 1 - \cos^2\Delta\chi + 2\cos\Delta\chi\cos\phi}{2\cos\Delta\chi\cos(\chi-\Delta\chi)\cos(\sin^{-1}[(\cos\Delta\chi\sin(\chi-\Delta\chi)])}\}$$

$$2\omega' = \cos^{-1}\{\frac{2\sin\theta\cos\Delta\chi\sin(\chi-\Delta\chi)\sin\beta + \cos 2\theta}{\cos(\sin^{-1}(2\sin\theta\cos\Delta\chi\sin(\chi-\Delta\chi))\cos\beta}\}$$

3.12

and

$$\Delta\chi = \sin^{-1}(2\sin(\varphi/2)\sin\varphi \qquad 3.13$$

φ is the angle between the diffracting plane and the sample mount normal, which is tilted at an angle χ with respect to the plane of the diffractometer, figure 3.26. The angle ϕ is the rotation about the normal to the plane of the sample mount to bring the diffracting plane into the plane of the diffractometer. ϕ is zero when the projection of φ on the plane of the diffractometer is at a maximum. If the tilt is small and φ is sufficiently large then the tilt can be aligned by rotation of ϕ, equation 3.13. $\Delta\chi$ is the angle of the scattering plane with respect to the surface normal to the plane of the diffractometer. The variation of the angles ω, $2\omega'$ with χ are given in figures 3.32a, b and c.

These derivations clearly relate to the condition that the reciprocal lattice point and the various divergences are negligible. These are not serious omissions but an extended reciprocal lattice point and an extended probe in the axial direction can cause problems in alignment. For example the diffraction plane normals of each mosaic block of a crystal or those of a layer and a substrate may not be able to be brought into the diffractometer plane at the same time. In which case this projection effect and subsequent errors, equations 3.12 and 3.13 can be significant. To overcome this another data collection method should be employed for the most precise measurements.

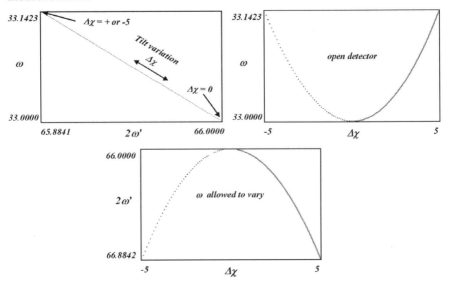

Figure 3.32. The variation in the measured peak positions as the tilt error is altered. The correctly aligned position for all three examples (a), (b) and (c) are when $\omega = 33^0$ and $2\omega' = 66^0$.

3.8.3.3. Three dimensional reciprocal space mapping

As discussed previously the reciprocal lattice points are three-dimensional, section 3.2, and therefore the axial divergence will integrate the intensity along this direction. Awareness of this third dimension is important since the conventional two-dimensional reciprocal space map is a projection onto the diffractometer plane. If the axial divergence is of the order of 0.5^0 and since the probe is purely a section of the Ewald sphere any intensity collected along this arc will be assigned to the wrong angle in the diffractometer plane. This is noticeable in these high-resolution measurements, especially when studying mosaic samples for example. An example will be given in the next chapter, but for now we will briefly describe the technique of three-dimensional reciprocal space mapping, Fewster and Andrew (1995, 1999), Fewster (1996). Restricting the axial divergence with crystals leads to a large reduction in intensity and is therefore not ideal, so slits are a good alternative, although the eventual probe still leads to a significant projection. However the view of reciprocal space can be changed dramatically.

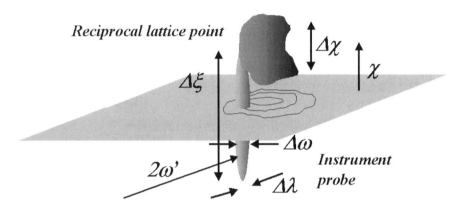

Figure 3.33. A schematic of how the data is collected for a three-dimensional reciprocal space map.

Since the divergence in the scattering plane is defined by crystal optics we only need to restrict the divergence in the axial direction. A 1mm slit at the monochromator exit and a 2mm slit in front of the detector is reasonable, using a point focus X-ray source. This will prevent scattering out of the plane of the diffractometer greater than about 0.09^0 from reaching the detector. We cannot actually define the divergence so simply, since as has been stressed throughout this book the whole process from source to detector has to be considered. Suppose the region of the sample that we are analysing (some feature that could be part of a mosaic block for example) is very small, say a few microns then this will define the divergence incident on the region. The actual axial divergence can therefore be exceedingly small for imperfect samples and this is how a significant amount of detail is observable. Although the detail is distributed over a large χ value this just leads to an effective magnification of the features of interest; this will become clear for the example given in figure 4.12. To understand the data collection method we should imagine the centre of the reciprocal space probe to be restricted to the plane of the diffractometer but can be moved around in this plane, figure 3.33. We will consider the height of the probe to be $\Delta\xi$ and the reciprocal lattice point to be tilted out of the diffractometer plane by χ and have a spread in this direction of $\Delta\chi$, which is a function of the shape and size of the reciprocal lattice point.

By collecting a series of reciprocal space maps, figure 3.31 and offsetting χ by a small amount (~0.2^0) after each map, we will create a three-dimensional array of intensity representing the distribution of scattering around the reciprocal lattice point in three dimensions. Of course the resolution is different in different directions, but it will create a far-improved data set for interpretation.

3.8.3.4. Applications of multiple crystal diffractometry

Let us consider some of the advantages of the multiple-crystal concept. Clearly we can analyse any material over very large regions of reciprocal space with very high resolution; i.e. the instrumental artefacts do not significantly smear the data. The probe can be moved throughout accessible regions of reciprocal space and the intensity can be mapped and from Chapter 2, the scattering process can be simulated and from the descriptions discussed above we can include all the instrumental artefacts. We can also extract information directly from the reciprocal space maps and these methods will be discussed in Chapter 4. However there are some additional benefits with this geometry, the angle 2ω' can be placed on an absolute scale and hence the lattice parameter can be measured absolutely to within a part per million and topography can be carried out on this instrument. The low divergence of the spatially large (millimetres in dimension) incident beam makes it suitable for topography. A photographic emulsion can be placed directly after the sample or after the analyser crystal. In the latter position specific areas in reciprocal space can be imaged (for example diffuse scattering) without the contribution of nearby strong scattering. These aspects will be discussed in Chapter 4 with examples.

The data collection time has not been discussed but basically as with all experiments this depends on the precision of the analysis and the problem to be solved. A typical intensity with a sealed X-ray source (2kW CuKα) arriving at the detector after passing through the monochromator and analyser (germanium crystals using the 220 scattering planes parallel to the surface) is close to 1,500,000 counts per second. This is enough to saturate most X-ray detectors. If the sample is close to perfect then it could scatter a very high proportion of this intensity and the hence the count rate during data collection can be very short depending on the dynamic range required. The lower limit, i.e. the residual count rate, is limited by the detector and or cosmic background and with careful setting of the energy window in the pulse height discriminator this can be reduced to about 0.1 counts per second. The dynamic range is therefore very large and allows very detailed analyses. Recovering some of the source divergence in the scattering plane with an X-ray multilayer mirror between the X-ray source and the monochromator can enhance the intensity further, section 3.6.4. The intensity magnification results from the decrease in the angular divergence exiting the mirror, that is the monochromator can accept more of the X-rays. The beam is obviously larger than it would be direct from the line focus source, but for many applications that is not a penalty. The improvement in intensity is greater than an order of magnitude, but the use of a line source instead of a point source increases the axial divergence although this can be controlled to a certain extent with a Soller slit, section 3.6.4. If the divergence of the monochromator is larger (i.e. asymmetrically cut crystals with low angle of incidence, figure 3.16a) then the intensity enhancement will be greater. The mirror in conjunction with the multiple-crystal diffractometer has advantages in rapid reciprocal space mapping, but the increased axial divergence introduces complications for precision lattice parameter determination and is unsuitable for topography. The advantages of being able to remove the mirror and revert to a point focus X-ray source depending on the application are very significant.

3.8.3.5. In-plane scattering in very high resolution

We discussed some of the basic theoretical concepts of in-plane scattering, section 2.9 and an experimental arrangement with slit based optics in section 3.7.3.2. There are advantages in carrying out in-plane experiments with high resolution, for example for measuring the precision lattice parameter of planes perpendicular to the surface or for studying the shape of the scattering close to Bragg peaks. The setting-up procedure is very similar to that described in section 3.7.3.2, with a narrow slit after the monochromator to limit the axial divergence for a good reflectometry profile. Initially to find the scattering from a plane normal to the surface an open window detector is ideal before invoking the analyser crystal. An indication of the sensitivity and resolution obtainable by this method is given in figure 3.34.

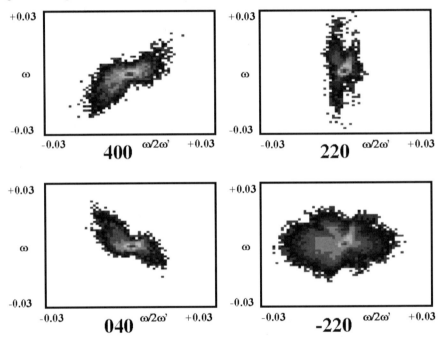

Figure 3.34. The diffuse scattering measured from a stack quantum dot layers, with the in-plane scattering geometry and the multiple-crystal diffractometer.

This particular example yielded the in-plane lattice parameters along four different directions and indicated that the whole structure had an average in-plane lattice parameter of that expected from the (001) GaAs substrate. Therefore no lattice relaxation was detected, however significant diffuse scattering can be seen and interpreted to give the asymmetry of InGaAs quantum dots grown on GaAs, Fewster (1999). By calculating the strains associated with randomly distributed dots in the plane but vertically correlated, the diffuse intensity can be calculated through equation 2.136 to yield the shape and composition Fewster, Holy and Andrew (2001). Clearly the dynamic range is sufficient to observe clear shapes and features within the diffuse scattering.

3.9. General conclusions

It should be clear from the discussions in this chapter that there are many possibilities for altering the instrument resolution to suit the problem to be solved. The discussion concerning the combination of components is far from complete, but an indication of the complexity and arguments for some arrangements have been given that should give an indication of the possibilities. Very simply the laboratory-based instrument is now becoming a very versatile tool for the analysis of most materials with intensity to saturate a detector and a resolution comparable to the most perfect materials available. The following chapter will discuss various examples and therefore indicate the performance and the type of analyses possible with some of these combinations.

Optimising the experiment to achieve the required data is always an important consideration and one aspect that is relevant to the discussion above is the size of the instrument function or probe. To obtain a complete set of data whether it is in the form of a reciprocal space map or a scattering profile the sampling volume for each data point should just touch each other or overlap. This will ensure no unexpected variations occur between data points. This is a particular problem with step scans when the diffractometer stops and measures the intensity at each position over a fixed time interval. This problem is overcome more successfully by continuously scanning and counting at the same time, the accumulated intensity is then associated with the angular mid-position of the count time interval. Hence a given count time determines the speed of the scan axis rotation. This is the most appropriate data collection procedure for diffractometers with encoders on the final axes. Stepper motors with encoders mounted on the motor axis are more suited to the step scan mode of data collection. Having encoders on the final axis is the ideal situation to determine the angular positions of the axes since it removes errors associated with slack gears and linkages. These are very important considerations for precision angular measurements. The two data collection strategies have been compared and no difference is observed provided that the step size does not exceed the probe dimensions.

References

Arndt, U W (1990) J Appl. Cryst. **23** 161
Arndt, U W, Duncumb, P, Long, J V P, Pina, L and Inneman, A (1998) J Appl. Cryst. **31** 733
Bartels, W J (1983) J Vac. Sci. Technol. B **1** 338
Compton, A H (1917) Phys. Rev. **10** 95
DuMond, J W M (1937) Phys. Rev. **52** 872
Fewster, P F (1996) Critical Reviews in Solid State and Materials Sciences, **22** pp.69-110.
Fewster, P F (1985) J Appl. Cryst. **18** 334
Fewster, P F (1999) J Mat. Sci.: Materials in Electronics **10** 175
Fewster, P F (1989) J Appl. Cryst. **22** 64
Fewster, P F (1999) Inst. Phys. Conf. Series No 164 pp197-206 IOP Plublishing Ltd: Bristol
Fewster, P F and Andrew. N L (1995) J Phys D **28** 451
Fewster, P F and Andrew, N L (1999) in *Defect and Microstructure Analysis by Diffraction.* IUCr monographs on Crystallography, Editors; R L Snyder et al, Oxford Univ. Press.
Fewster, P F, Holy, V and Andrew, N L (2001) presented at the ICMAT, Singapore, to be published.
Hamilton, W C (1974) *International Tables for Crystallography* **IV** 280
Hart, M (1981) J Cryst. Growth **55** 409
Iida, A and Kohra, K (1979) Phys. Stat. Solidi A **51** 533
Komakhov, M A and Komarov, F F (1990) Physics Reports **191** 289

CHAPTER 4

A PRACTICAL GUIDE TO THE EVALUATION OF STRUCTURAL PARAMETERS

4.1. General considerations

So far we have built an understanding of the structural properties of our sample, the scattering theories and how the instrument collects the data. In this chapter some examples of typical analyses will be given. This will not be comprehensive but should give examples of the methods that have been developed. Firstly we should reiterate some of the discussion so far.

The sample is defined in terms of several structural properties, i.e. shape, composition, crystallinity, etc., and these properties determine the appropriate experimental arrangement as well as the parameter of interest. It is also very important to recognise the interaction between these parameters, so in extracting a single parameter value, it is important to be aware of the other parameters and their influence on the scattering. The type of experiment is therefore greatly influenced by the knowledge that we already have concerning the sample.

The scattering theory to choose will depend on the material properties and the precision required. In chapter 2 we covered a range of scattering theories, all of which have assumptions and must therefore only be used beyond these boundaries with care. However it is clear that we do have theoretical methods to cover from the most imperfect to the most perfect crystal samples. The depth of information that can be obtained is quite considerable.

In chapter 3 we discussed the instrumentation for analysing samples. It is crucial to recognise that we cannot isolate the influences of the sample and instrument without very good knowledge of one or the other. Basically if we wish to know detailed structural information about the sample then we require a very well defined and predictable X-ray probe. If we know considerable detail about the sample then we can use cruder experimental techniques. It is therefore important to emphasise that the instrument should be versatile so that it can be adapted to the problem of interest. We cannot simply define an instrument function and a sample response and form a correlation of the two influences. The reason for this is that the sample may not make full use of the X-ray probe instrument function. This can lead to profile widths being narrower than the instrument function, for example.

The conclusion of all this is that the diffraction experiment should be considered as a whole; X-ray source, incident beam conditioning, scattering from the sample and detection with scattered beam conditioning. Of course as we understand more about the sample problem to be tackled, the influence of other structural parameters and the performance of the appropriate instrument configuration we can simplify the analysis considerably. The following examples presented use a range of techniques all having their merits, some for speed, some for precision, etc., depending on the level of assumptions that are made. There are further examples on a whole range of material types discussed in Fewster (1996).

4.2. General principles

From our understanding of reciprocal space we can see that we have a large range of scattering to analyse in three dimensions. Depending on the nature of the sample, some of this will be inaccessible. The choice of wavelength and diffractometer geometry will also limit the accessibility, but in general the information will be over defined and much of the analyses can be carried out using a few reflections. As discussed in chapter 1, semiconductors that are manufactured will have details of their structure known but not to the precision necessary, it is therefore only the fine detail of exact composition, thickness and perfection that is often required. In this case the analysis could be restricted to the region close to a single reflection. If the structure is composed of several layers and some critical thickness is reached, so that the normally assumed strained state has relaxed, then we require more reflections to analyse this strain state to obtain compositions, etc. If the relaxation has progressed to the stage of breaking the sample into small mosaic blocks then these details require reciprocal space mapping to analyse their shape and so on.

Clearly with the analytical tools in X-ray methods we can produce an excellent understanding of our sample, by combining reciprocal space mapping, topography and simulation. For very complex structures often the number of parameters can be too large for a complete 'all in one analysis' and the dominant parameters need to be determined first. From this we can carry out further experiments to obtain a complete picture of our sample, by using several reflections (some are more sensitive to different properties). It is important to establish uniqueness as in any analysis and in general X-ray experiments use a very small fraction of the total available data and hence there are plenty of opportunities to quickly test the proposed model. Generally all the parameters are correlated in the scattering pattern and therefore the full simulation of the sample with the most exacting theory is one of the most rigorous tests we can conduct. As mentioned above X-rays are very sensitive to deviations from perfection, which is a significant advantage, however this can make interpretation more difficult. It is the purpose of this chapter to help with this through examples. It should also become clear that the resolution of the probing instrument should be carefully matched to the problem and therefore there will be significant referral to the sections in Chapter 3. Once the understanding becomes more complete then there are many assumptions that we can make and the analysis can be exceedingly fast. We shall start with examples that illustrate the use of direct interpretation and how this can also lead to errors. Various approaches to some of these measurements have been previously discussed in Fewster (1996), it is more the purpose here to concentrate on the preferred method (simplest or most accurate, etc.), and describe the chosen experimental conditions.

4.3. Analysis of bulk semiconductor materials

Bulk semiconductors are primarily used for substrate material or ion implantation and their quality is often the basis for good epitaxial growth. The important parameters associated with bulk material will therefore relate to the general "quality", the chemical composition, the absolute lattice parameter throughout the sample (hence the state of strain), the orientation

174 X-RAY SCATTERING FROM SEMICONDUCTORS

of the crystallographic planes with respect to the surface and the surface quality.

4.3.1. Orientation

Orientation is fundamental to all subsequent experimental techniques since this defines the region for more detailed analyses. Generally a wafer ready for epitaxy is orientated and specified by the supplier and therefore this is often assumed. However details of the directions associated with [110] and [1-10] for a [001] surface of GaAs, for example are rarely presented.

4.3.1.1. Surface orientation – the Laue method

The Laue method is perhaps the simplest experiment in X-ray scattering. It consists of simply placing the sample in an X-ray beam and collecting the scattered beams on a photographic plate. The geometry of the back-reflection Laue method and a convenient precision camera is given in figure 4.1.

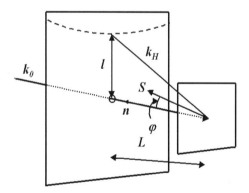

Figure 4.1. The geometry of the back-reflection Laue method. The X-rays pass through the film to the sample and are then scattered back onto the film.

Ideally the mechanical design should be such that removing and replacing of the sample holder maintains a precision for obtaining misorientations to within $0.05°$. A typical scattering pattern is given in figure 4.2, for a GaAs wafer taken with a crystal to film distance of 3cm.

The symmetry of the pattern is at first sight clearly 4-fold and any distortion of this pattern such that the centre of symmetry is displaced from the centre of the pattern will relate to the misalignment of the crystallographic planes with respect to the surface. The symmetry of the pattern relates to the symmetry of the surface normal. Methods of interpretation for general orientations have been given by (Cullity, 1978, Fewster, 1984, Laugier and Filhol, 1983). These methods work well for high symmetry space groups and become progressively difficult with lower symmetries.

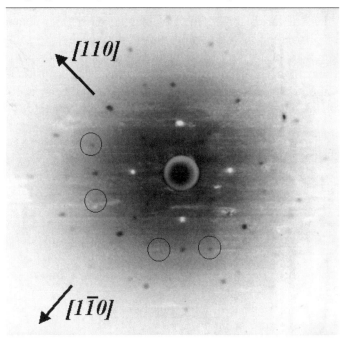

Figure 4.2. The Laue back-reflection image from a (001) GaAs sample. The symmetry of the pattern is close to 4-fold however the small intensity differences (circled, strong pairs indicate the [110] direction) show up the 4-fold inversion axis that has 2-fold symmetry.

The pattern in figure 4.2 is taken with Polaroid film with an exposure of 5 minutes using a 0.8kW X-ray Cu source. The preferred X-ray source anode for this work is tungsten since it gives a broad region of white radiation. The exciting voltage across the X-ray tube is limited to 20kV to reduce the fluorescent yield that increases with higher energies, and the tube current is increased to boost the flux from the source. The principle of the technique is illustrated in figure 4.1. The full spectral distribution of the source is used and for a stationary sample each crystal plane orientation with respect to the surface plane, φ, and inter-planar spacing, d, will scatter if it satisfies the following condition based on the Bragg equation

$$\varphi = \frac{\pi}{2} - \sin^{-1}\left\{\frac{\lambda}{2d}\right\} \qquad 4.1.$$

Therefore with a large spread in wavelengths each inter-planar spacing accessible will scatter in a direction that is defined by the sample itself. However the intensity of the scattering is very complex and each maximum can have contributions from many wavelengths. We can determine the misorientation of anyone of these crystal planes from the surface normal by determining the length l in figure 4.1, and from simple geometry

$$\varphi = \frac{1}{2}\tan^{-1}\left\{\frac{l}{L}\right\} \qquad 4.2$$

Suppose now that we want to measure the misorientation of the surface with respect to the [001] direction in figure 4.2, then we have to use symmetry-related maxima to define the centre of the pattern. Just using the intersection of straight lines between these maxima is adequate for small displacements where the central maximum is hidden, although strictly they lie on hyperbolae.

The above approach will achieve accuracies of the order of $0.3°$ with care. To improve this accuracy considerably we can move the sample further away from the film (increase L, equation 4.2.), and also account for any geometrical alignment errors in the camera, Fewster (1984). Suppose that we take two photographs of the Laue pattern, the second after rotating the sample through $180°$, then the pattern should rotate about its surface normal that should coincide with the camera centre if it is perfectly aligned. Take the measured centres of the two patterns as having co-ordinates (x_{C1}, y_{C1}) and (x_{C2}, y_{C2}) then the true length l is simply

$$l = \left[\left\{\frac{(x_{C1}+x_{C2})}{2}\right\}^2 + \left\{\frac{y_{C1}+y_{C2}}{2}\right\}^2\right]^{1/2} \qquad 4.3$$

where x and y represent the distances horizontally and vertically from the perceived camera centre, C for the two exposures, 1 and 2. This also gives a method of aligning the camera.

The errors associated with this method are discussed in Fewster (1984) and for a 5cm crystal film distance these permit measurements to within $0.1°$ and at 10 cm it is possible to achieve measurements to within $0.05°$.

4.3.1.2. Determining the orientation by diffractometry

In this section we shall discuss two methods where the choice is determined by the instrumentation available. In the first method we require a well defined incident beam (e.g. 2-crystal 4-reflection monochromator or a double pinhole slit arrangement) and precision ϕ and ω axes (i.e. the ϕ axis should be normal to the sample mounting plate. The second method relies on multiple crystal methods (i.e. the multiple crystal diffractometer, section 3.8.3) and precision axes (both ω and 2ω'). The principles of the two methods are quite different and the latter is considerably more precise than the former. Any method relies on defining the surface plane very precisely and relating this in some way to the crystallographic planes.

4.3.1.2.2. Monochromator and open detector method

This is most suited to obtaining approximate values of misorientation. The sample is placed on the sample mount so that the sample surface is accurately parallel to the mounting plate. The precision of this basically defines the precision achievable. The larger the sample wafer the smaller the error, also if the sample is not parallel sided then the sample should be mounted with its surface facing the surface mount.

A suitable reflection should be found with the detector in the wide-open position, figure 4.3a, at the Bragg condition $\omega' = \theta$. The incident angle now corresponds to the Bragg angle (ignoring refractive index corrections) plus the crystal to surface plane angle. By rotating the ϕ axis the intensity will start to decrease unless the misorientation is zero. If the rotation is a full 180^0 then the angle through which ω must be rotated to recover this intensity is twice the angle between the sample surface and the scattering planes, φ.

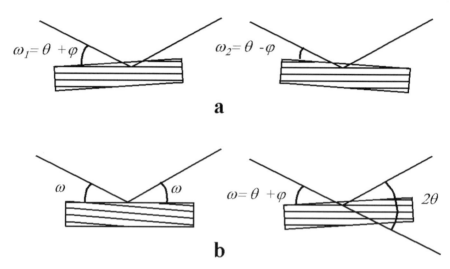

Figure 4.3. Two methods for determining the orientation of the crystallographic planes with respect to the surface plane. (a) relies on the precision rotation of the surface plane through 180^0 between measurements and (b) works on the principle of precision incident beam and scattered beam measurement.

$$\varphi = (\omega_1 - \omega_2) \qquad 4.4$$

This can be repeated for several azimuths until a clear sinusoidal variation is observed. The maximum in this oscillation is the largest tilt direction and the azimuth at which it occurs. If however the measurements are carried out at two orthogonal azimuths, i.e. rocking curves at 0^0 and 180^0 to obtain φ_0 and at 90^0 and 270^0 to obtain φ_{90}, then the maximum φ is given by

$$\varphi_{max} = \tan^{-1}\left\{\sqrt{\tan^2\varphi_0 + \tan^2\varphi_{90}}\right\} \qquad 4.5$$

and this occurs at the azimuth given by

$$\phi_{max} = \tan^{-1}\left\{\frac{\tan\varphi_{90}}{\tan\varphi_0}\right\} \qquad 4.6$$

4.3.1.2.3. Multiple crystal diffractometer method

The difficulties in referencing the sample surface to sample mounting plate are removed with this approach. From section 2.3.1 we could see that at low incident angles the incident beam is specularly reflected and therefore the scattering angle, $2\omega'$, is precisely twice the incident beam angle ω. This is achieved by placing the analyser assembly at some angle just above the critical angle and scanning ω until the maximum intensity is found. We now have ω and $2\omega'$ on an absolute scale.

We now find a suitable reflection and rotate the ω and $2\omega'$ axes until the scattering condition is maximised, $\omega' = \theta$. Initially the reflection will be found with no analyser, but with a good precision goniometer moving between the two does not loose this reference. The ω angle can be determined from the peak position with the open detector and once set at this maximum in the intensity the analyser assembly can be invoked and scanned to determined the scattering angle, figure 4.3b. If $2\omega = 2\omega'$ then the surface and crystallographic planes are parallel to within the accuracy of the instrument resolution, i.e. a few arc seconds. If $2\omega \neq 2\omega'$ then the misorientation is given by

$$\varphi = (\omega - \theta) \qquad 4.7$$

The maximum tilt is now given by equation 4.5 and its azimuthal position by equation 4.6. The precision of this method is exceedingly high and only relies on the precision of the main ω and $2\omega'$ axes and judgement of the peak positions that are defined by the intrinsic scattering width. The most precise determination of the peak position is achieved by performing a very small reciprocal space map around the peak with the minimum step size available.

4.3.1.3. Determining polar directions

From Chapter 1 we could see that the zinc blende structure that is characteristic of GaAs, InP based materials, etc., are not centrosymmetric. This is also evident in the manner in which the bonding at the free surfaces of these materials reconstruct and this will clearly influence the epitaxial growth on these materials. Some of these influences may be minor for [001] orientations but for [111] orientations this can become very pronounced, where the bonding normal to the surface is either Ga to As or As to Ga for example.

To determine these details we have to consider more subtle influences on the scattering patterns. In Chapter 2 we discussed the structure factor, which is the phase sum of the scattering factors of the atoms, i.e. its value depends on the distribution of atoms in the unit cell. From equation 2.22

$$F_{hkl} = \sum_{j=1}^{N} f_j e^{2\pi i(hx+ky+lz)} \qquad 4.8$$

Now if the scattering from each atom involves no absorption (i.e. f_j is assumed real), then the structure factor associated with the planes (hkl) and $(-h-k-l)$ will be complex conjugates of each other. If we take the simple kinematical description of intensity ($I_{hkl} = F_{hkl}F^*_{hkl}$) the intensities will be identical. However f_j is not wholly real because the presence of absorption. The degree of absorption depends on how close the X-ray frequency matches the resonant frequency of one or more of the sample atoms. The imaginary contribution to the scattering factor (f'') will always retard the phase of the scattered wave by

$$\Phi = \tan^{-1}\left(\frac{f''}{f+f'}\right) \qquad 4.9$$

Since the contribution (i.e. phase retardation) is the same for F_{hkl} and F_{-h-k-l} they are no longer complex conjugates and the intensities of reflections from the (hkl) and the $(-h-k-l)$ planes are no longer equal. We can use this difference to compare X-ray scattering from (hkl) and $(-h-k-l)$ crystal planes.

Chapter 4 A Practical Guide to the Evaluation of Structural Parameters 181

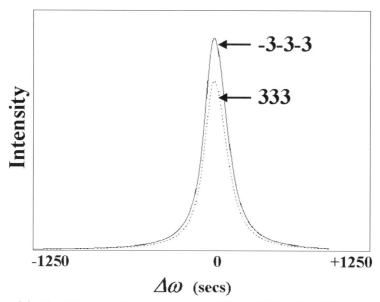

Figure 4.4. The difference in the scattering strength from (333) and (-3-3-3) crystallographic planes from CdTe with CrKα radiation.

Figure 4.4 gives the scattering profiles and the integrated intensities from the 333 and -3-3-3 reflections of CdTe using CrKα radiation. The choice of wavelength here is to enhance the values of f' and f'' for both Cd and Te and emphasise the effect. We can see quite clearly that the intensity difference is very significant and easily measurable. Since we may wish to determine the surface to grow on for epitaxy (some material systems produce far better growth on some surfaces than others) we have to remount the wafer. When comparing opposite sides of the same wafer, the incident beam should be monitored and all the scattered intensity measured (in case the sample is mosaic or bent) or an insensitive reflection as an internal reference should be used. A suitable choice in this case is the *222* reflection, which has the same intensity as the *-2-2-2* reflection (Fewster and Whiffin 1983).

Another important polar direction that can influence the over growth of epitaxial layers is that of distinguishing directions in the surface plane, although the surface plane and its reverse are identical. The principal directions, i.e. *{110}* type, can be observed directly from the Laue pattern of figure 4.2, however *[110]* and *[1-10]* are not the same in non-centrosymmetric crystals. We indicated earlier that the Laue method is multi-wavelength and therefore we can make use of this by finding a reflection with an appropriate *d*-spacing that selects a wavelength between the Ga and As absorption edges for example, equation 4.1. Firstly select a wavelength range that enhances the effect and then find a range of surface to diffraction plane angles, φ, that can be observed (depending on the film size) and determine appropriate *d* values and hence *hkl* that will show this. For the *(001)* orientated GaAs example of figure 4.2 a reliable reflection is the *139* type, that occurs in eight positions in the pattern and scatters with a wavelength of 1.1183Å. The reflection with indices *139, -1-39, 319, -3-19* are all stronger than the *–139, 1-39, 3-19, -3,19* reflections. So once we have established the reflection to use any wafer can be quickly orientated in this way, (Fewster, 1991a; Schiller 1988).

4.3.2. Revealing the mosaic structure in a bulk sample

As we could see from Chapter 2 the intrinsic scattering profiles are very narrow for nearly perfect materials and any regions that have different orientations from the average will scatter with different strengths for a fixed incident angle. We can consider a mosaic block as a region of perfect crystal surrounded by low angle boundaries. A mosaic block structure can be formed in a crystal from the convection currents close to the growing liquid-solid interface. By their very nature the blocks will be tilted and if they are very small they could create diffraction broadening effects. Mosaic crystals can therefore create dramatic changes to the scattering.

4.3.2.1. Mosaic samples with large tilts

The first example we shall consider is a sample of $LiNbO_3$ that has large mosaic crystal blocks (few mm) with significant relative misorientations ($>0.05^0$). The simplest method for analysing samples similar to this is by low-resolution topography. The principle of the method is illustrated in figure 4.5a. The divergent X-ray source relies on the different orientation of the sample scattering planes to select the appropriate incident beam direction, and since the scattering angle is the same for each block the scattered beam leaving the sample is rotated by the misorientation angle.

Hence each image from each mosaic block is displaced, figure 4.5b. The degree of overlap or the size of the gap along with knowledge of the sample to film distance will give the angle of relative tilt. The resolution and nature of the geometry leads to the benefit that this measurement can be made perpendicular and parallel (and anywhere in between) to the scattering plane.

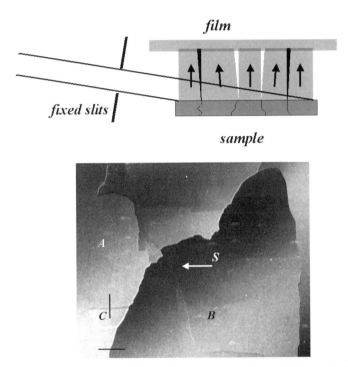

Figure 4.5. (a) The principle of the Berg-Barrett topographic method for analysing mosaic crystals. (b) The topograph of a LiNbO$_3$ crystal showing the black/ white contrast at the mosaic block boundaries. The misorientations measured were of the order of 10 minutes of arc.

The resolution is defined by the intrinsic scattering width of the sample since this method relies on the self-selection of the incident divergence. Hence, although this method is termed low-resolution, dislocations can be observed when the scattering widths of the perfect regions are very small, Fewster (1991a).

4.3.2.2. High resolution scanning methods (Lang method)

When the misorientation between the mosaic blocks is very small then the angular sensitivity has to be increased by reducing the angular divergence of the incident beam. To maintain the small divergence yet image large areas it is best to scan the sample back and forth through a restricted divergent beam. This method can be successfully used with a Lang camera in reflection mode using a narrow slit (~200μm) in the scattering plane and the dimensions of the sample in the plane normal to this. A distant X-ray focus is used to achieve a divergence of ~ 0.002^0 (~10"arc), figure 4.6a. This method is not without its problems associated with bent samples and long exposure times. The image will in general consist of the medium strength scatter associated with the mosaic blocks and more intense regions associated with the surrounding defects (these extract more intensity out of the divergent beam). However the differing contrast between the mosaic blocks will give an indication of orientations greater than the divergence of the incident beam.

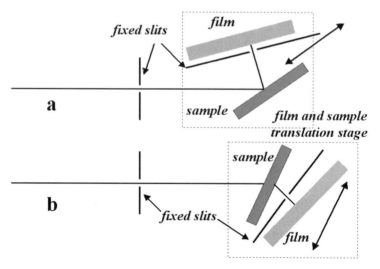

Figure 4.6. Two arrangements of the Lang traverse topography method; (a) reflection and (b) transmission settings.

Chapter 4 A Practical Guide to the Evaluation of Structural Parameters

Figure 4.7. A Lang transmission topograph of a 150mm diameter Si sample taken with MoKα radiation. The substrate was heavily boron doped with a lighter doped layer. The presence of dislocations, (bottom and top right) and swirl (rings) are all evident in this image.

The depth of penetration of the incident beam close to the Bragg condition is severely limited by the scattering process and therefore these reflection techniques will only give information within the top 10μm or so. Transmission topography, figure 4.6b, gives a closer representation of a bulk substrate quality however the absorption would seem to be too great to obtain anything worthwhile. For silicon the absorption of MoKα is not inhibiting and we can obtain reasonably clear images without difficulty. The value of μt for a 700μm wafer for example is ~1. From section 2.10 we can see that the process described above dominates the contrast in the image, i.e. the defects extract more divergence out of the incident beam and create a more intense image than the perfect regions, figure 4.7. However GaAs is highly absorbing for both CuKα and MoKα radiations and therefore need to be thinned to ~40μm to achieve a similar contrast. In section 2.10 we indicated that the absorption for each wave-field and polarisation differs and our notion of a general absorption value is not valid.

We can therefore make use of this feature for imaging unthinned wafers where the signal is very weak, but certainly measurable. The sample is set and aligned in the normal way however only the single wave will emerge from the exit surface with a very small angular spread (i.e. the scattering peak is weak and sharp and therefore more difficult to align). The contrast is formed in a very different way since the transmission relies on the perfect regularity of the atoms in the sample. Any deviations from this perfect alignment will almost certainly mean the X-rays will be deviated and absorbed resulting in missing intensity. The contrast is reversed from that of the low μt samples, figure 4.8. These methods do reveal useful information but are rather slow and therefore we will consider a somewhat faster method.

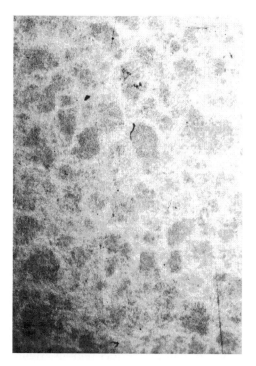

Figure 4.8. An anomalous transmission Lang topograph from GaAs with a $\mu t \sim 18$ for MoKα showing the loss of intensity where there are a high number of defects. The GaAs is semi-insulating and full of mosaic blocks.

4.3.2.3. Multiple crystal methods for revealing mosaic blocks

We shall consider a semi-insulating GaAs wafer and examine the mosaic structure using reciprocal space mapping and topography, Fewster (1991b). If we start by undertaking a high resolution rocking curve (the incident beam is well defined, the detector is wide open and the incident angle on the sample varied, ω) then we obtain a profile that is very broad, many times that of the intrinsic calculated profile, figure 4.9.

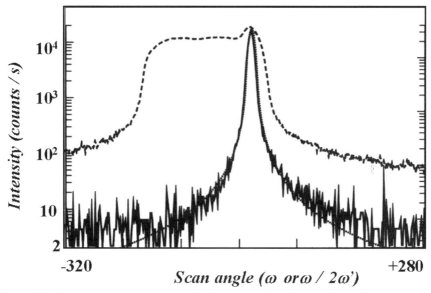

Figure 4.9. The scattering from a mosaic GaAs sample; dashed curve the profile with an open detector (ω scan), the continuous curve is that obtained from the multiple-crystal diffractometer ($\omega/2\omega'$) compared with that expect theoretically, dotted line.

At this stage we cannot make too many conclusions since we are mixing contributions of bend, mosaic block orientation, strain variations and the intrinsic scattering all into this one profile. The introduction of an analyser crystal, using the geometry of figure 3.27 allows us to scan along the plane normal direction that is sensitive to strain, figure 3.30b, immediately it becomes clear that this profile matches that expected from the intrinsic profile for a perfect crystal. The major contributing factor to the width of the open detector scan is therefore normal to this direction. If we now centre the probe on the peak of figure 4.9, using the configuration of figure 3.27, and scan in ω and keep the detector at the scattering angle $2\omega'$ (= 2θ) of the peak, then we can measure the orientation spread over the area of the beam on the sample, figure 4.10. It is important to realise that any measure of angle is a projected angle onto the plane of the diffractometer, therefore we can only use this angle quantitatively if we assume the orientation distribution is isotropic and the sampled region is representative. These latter points are easily confirmed with experiments performed at different ϕ azimuths.

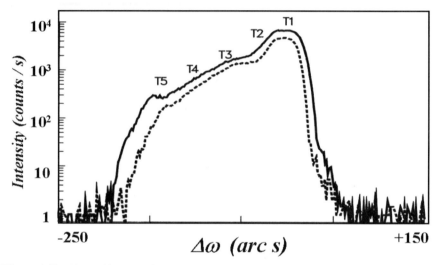

Figure 4.10. The rocking scan for a mosaic structure with a multiple-crystal diffractometer. The dashed curve is a repeat scan with a slit reducing the sampled area in the scattering plane, the width does not scale with this reduction and therefore is not due to curvature.

A more complete understanding will be obtained by collecting a high-resolution diffraction space map (high-resolution 2C4R monochromator and 1C3R analyser, section 3.8.3). The map is obtained by undertaking a series of scans by driving the $2\omega'$ and ω axes in a 2:1 ratio and then offsetting ω by a small amount and repeating, figure 4.11. The diffraction space map contains all the information of the three scans, figure 4.9 and 4.10. With this map it is now possible to see whether there are strain variations associated with different orientations. It is again important to recognise that a map is a projection of a small section of the Ewald sphere onto the plane of the diffractometer. Therefore any mosaic blocks with scattering planes inclined to the main axis of the diffractometer will give an error in its ω and $2\omega'$ position and therefore there will exist an uncertainty in the value of the derived strain. The extent of the error introduced is covered in the sections 3.8.3.2 and 4.3.4 on lattice parameter determination.

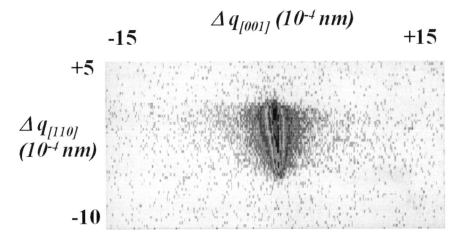

Figure 4.11. The reciprocal space map of the mosaic structure described in figures 4.9 and 4.10. The dashed line is the same scan with the beam restricted in the scattering plane with slits. The similarity of the profiles indicates that the shape is not primarily due to sample curvature.

To overcome all these projection errors Fewster and Andrew (1995a) have introduced the concept of three-dimensional reciprocal space mapping. In this procedure the axial divergence is heavily reduced at the monochromator and analyser, which reduces the intensity, but the diffraction space probe is contained close to the plane of the diffractometer. The projection effects are severely reduced and the data is collected as a series of reciprocal space maps at different tilt values, section 3.8.3.3. Hence only scattering that is contained in the bounds of the diffraction space probe, still at very high angular space resolution, and does not deviate from ~0.2° of the plane of the diffractometer can be accepted. The distribution of intensity will then appear as in figure 4.12 for a mosaic GaAs substrate with an AlGaAs layer on top from which the true strain associated with each mosaic block can be determined. This method has also been used for analysis of polycrystalline and single crystal enzyme structures, Fewster (1996).

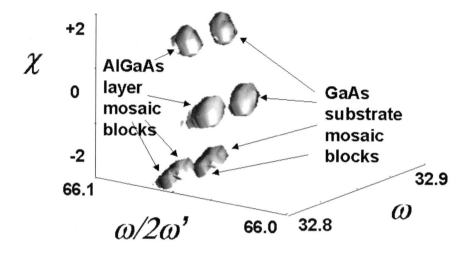

Figure 4.12. A three-dimensional reciprocal space map showing the orientation relationship of an AlGaAs layer to that of the GaAs substrate. Note also how the shape is mimicked from substrate to layer.

Of course these measurements give some quantitative values of the degree of strain and the orientation spread, etc., however it does not reveal the dimension of the mosaic blocks. For this particular sample the mosaic blocks are large ~0.7mm in diameter and they can be separately imaged by topography. In general the probing beam although very small in diffraction space is large in real space, approximately 8mm x 1mm. Therefore the data collected so far is a contribution from a large region on the sample. However within the probed area we have regions of varying contribution. By placing a fine-grain photographic emulsion with very small grains in the scattered beam from the sample we can observe contrast on a scale comparable to the developed grain size of a few microns. Individual images can be observed if their relative orientation is greater than the divergence of the incident beam. The spatial resolution of this method differs in the axial and scattering planes. In the scattering plane the resolution is defined by the divergence of the incident beam and the wavelength spread, see Chapter 3, which have a negligible influence on the spatial spreading of the image. The scattering plane resolution is therefore limited by the developed grain size of the emulsion. The axial resolution is influenced by the wavelength spread along this direction and the axial divergence passed by the monochromator as well as the overall geometry. The wavelength spread is small but this effectively allows the beam from the point source to pass and the spatial resolution is dominated by the geometry. The resolution is therefore simply given by

$$R_{axial} = \frac{film...to...sample...length}{source...to...sample...length}(source...height) \qquad 4.10$$

In practice it is difficult to place the film closer than about 10mm from the sample and using a typical source to sample distance of 300mm and point focus (0.4mm high), the axial resolution is about 13 microns. This compares with 1 micron in the scattering plane. The shape of the defect images will be smeared to this extend by the instrumental resolution.

Figure 4.13. X-ray reflection topographs at positions denoted by figure 4.10 of a GaAs mosaic crystal. Defects at the edge of each block are just visible and a topograph taken whilst rocking the sample creates a continuous image. The topographic emulsion was placed immediately after the sample.

Take figure 4.10, where we know that each data point comes from an object that satisfies this orientation and strain value. We have assumed that the strain broadening is small, figure 4.9 and therefore placing a photographic emulsion immediately after the sample, figure 3.27 will create an extended probe (described in figure 3.30a as an open detector probe). However because of the nature of the shape of the reciprocal lattice point, figure 4.11, the dominant scattering image will relate to the orientation differences. Hence a series of topographic images at various rotations in ω will image the contributing feature. In this case the mosaic blocks are large and do not create diffraction broadening along ω, but each image is of a mosaic block surrounded by defects accommodating the small tilt boundary. In this case the relative tilts are greater than the divergence of the incident beam (~5"arc) and therefore each block can be imaged separately, figure 4.13. Because each image is obtained after sliding the film cassette perpendicular to the ω axis the relative positions of the blocks can be related to sample (provided that the axial plane tilt is small). This type of topograph can be obtained at the peak maxima and therefore the exposures can be very short (a few minutes).

4.3.3. Characterising the surface quality

It is clear that the topography methods discussed above rely on relative contrast of the scattering from the perfect regions and the defects. A significant advantage of reciprocal space mapping with a very small scattering probe is that we can separate the scattering close to maximum from a nearly perfect sample into distinct features: Bragg scattering, surface truncation rod or dynamical streak and diffuse scattering, figure 4.14. The Bragg scattering comes from the bulk of the sample and is effectively a point in reciprocal space, i.e. the Fourier transform of an infinitely large three-dimensional object. The surface truncation rod is related to the surface, i.e. the Fourier transform of an extended two-dimensional plane. The diffuse scattering on the other hand comes from the defects existing in the material, the shape of the scattering could be related to the defect type however there can be many contributors to this. From the topography methods above, the images are collected that will include all these contributions. Suppose now we centre our small reciprocal space probe on the diffuse scattering and form an image of the scattering over the probed area using a photographic emulsion immediately after the analyser crystal, figure 3.27. With this arrangement we are obtaining an image of just the defects.

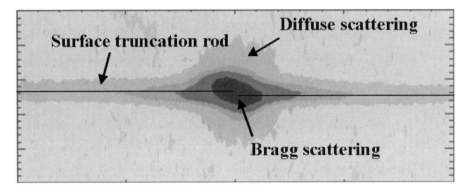

Figure 4.14. The main features of the scattering from a nearly perfect bulk crystal (004 reflection of GaAs).

The defects surrounding mosaic blocks can be imaged rather easily in this way and also surface scratches, associated with surface preparation techniques show very easily. Different regions of the diffuse scattering can be associated with different defects, Fewster and Andrew (1993a), however by placing the probe very close to the Bragg peak the intensity can be enhanced and most of the defects will contribute to the image, figure 4.14. The exposure times can be as short as ~1h (using a 2kW CuKα point focus and L4 nuclear emulsion plates) depending on the number of defects contributing to the image. If few defects are present then the intensity recorded at the detector will be concentrated into small grains and the exposure is shorter, whereas distributed defects will lengthen the exposure time for a given detected count-rate.

Occasionally the surface streak can exhibit some rather strange wiggles. These wiggles are only observed with very high resolution and also relate to the surface structure. This can sometimes be most obvious with a reciprocal space map (small undulations) or a multiple crystal scan normal to the surface plane (the intensity either side of the Bragg peak has a different variation), figure 4.15. This variation can be modelled by assuming the surface is undulating in a rather asymmetric form, figure 4.16.

194 X-RAY SCATTERING FROM SEMICONDUCTORS

This example should indicate that the surface can be characterised at all levels and with a fairly complete theoretical model we can reconstruct the surface.

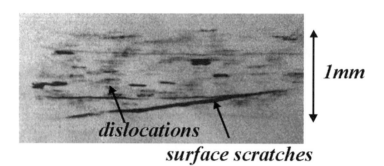

Figure 4.15. A diffuse scattering topograph of a bulk sample crystal indicating surface damage and dislocations. The exposure time was 4 hours.

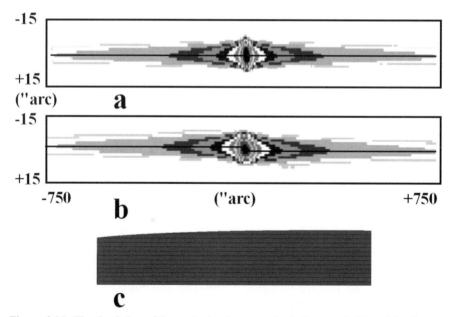

Figure 4.16. The simulation of the scattering from a perfectly flat sample (a) and that from a sample with a gradual small undulating surface (b). In this simulation 30% of the surface is perfectly flat with a gradual undulation leading to a maximum of $2.8°$ misorientation over 2% of the surface (c). Experimentally this effect is evident in figure 4.14.

4.3.4. Measuring the absolute interatomic spacing in semiconductor materials:

So far we have concentrated on the general aspects of quality, but have not discussed aspects of strain within the sample. The strain variation can be determined over the scattering volume with an $\omega/2\omega'$ scan using the multiple crystal diffractometer as in the previous example. The strain variation will then be quantified by comparison with the calculated profile assuming the material to be perfect. However the absolute interplanar spacing in a sample is sometimes very important to know since this can form an internal reference for composition analysis or for determining the true state of layer strain or lattice parameter relaxation.

The assumptions and requirements of any measurement depend on the quality of the sample and the standard to which it is related. We will assume that the wavelength of the peak of the CuKα spectral line is known, or is at least a transportable standard. Since this depends on a quantised transition in a Cu anode of the X-ray tube this is essentially invariant although we may argue on the relative value to that of the defined standard of length internationally. Standard materials are very difficult to achieve due to differences in processing, surface preparation, mounting strains, etc. Another requirement is the homogeneity of the sample if the method relies on two measurements as in the method proposed by Bond (1960). A single measurement is therefore ideal. The method described here is that of Fewster and Andrew (1995).

The principle is simple in that the incident beam and accepted scattered beam directions can be defined very precisely using the multiple crystal diffractometer, figure 3.27. The incident beam direction is precisely defined by scanning the analyser / detector assembly through the path of the beam from the monochromator. The width of this profile is very narrow, figure 3.28. The centre of this peak defines the condition $2\omega' = 0$. Scattering from the sample will now be on an absolute angular scale provided that it is confined to the plane normal to the analyser / detector axis and the integrity of the angular movement is precise. The sample does not have to be accurately centred on the diffractometer nor does the incident beam have to pass through the centre of the goniometer axis, although in general this makes the whole process easier since this prevents the beam from being displaced spatially. This reduces the main errors that are very difficult to compensate for or know with certainty, figure 4.17.

196 X-RAY SCATTERING FROM SEMICONDUCTORS

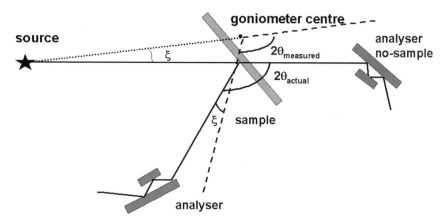

Figure 4.17. The principle of the method of determining the interplanar spacing with a single measurement. The zero scattering angle is determined from the direct beam, the X-rays do not necessarily have to pass through the goniometer centre nor does the sample have to be mounted precisely to achieve 1ppm precision. Dashed line is the assumed beam path.

The only alignment necessary to ensure that the scattering planes are parallel to the main axes, section 3.8.3.2. The most precise alignment procedure used is to initially rotate the sample in ϕ so that the plane of interest, with an angle of φ to the surface normal is maximised in the plane of the diffractometer. The sample should now be brought into the beam so that it cuts the incident beam in half. The sample should be scanned in ω to ensure the surface is parallel to the beam. The maximum in this intensity corresponds to the $\omega = 0$ position. The sample and detector should then be driven to their respective positions to capture the scattering of interest. Then for a given ω value, where there is intensity, the χ axis should be scanned and the aligned position corresponds to the midpoint of the profile. This procedure is done without the analyser crystal and just an open detector. If the profile is double peaked, this is generally the case unless the alignment is very close, then the intensity can be recovered by driving the ω axis to a lower value from this midpoint, figure 3.32a. The alignment is much more reliable when the axial divergence is restricted as much as possible with a slit since this will decrease the spread in acceptable tilt values and sharpen the profile and additional contributions from differently tilted regions in imperfect samples. The intensity is peaked with the ω axis.

The analyser should now be inserted (apart from driving the instrument axes there should be no touching of this unit to maintain the integrity of the measurement. The 2ω' axis can now be scanned to find the maximum intensity value; this now gives the equivalent of a cross hair on the reciprocal space map. To determine the 2ω' angle precisely a scan along the scattering vector should be performed or a reciprocal space map collected along 2ω' / ω. As before the most precise method of all is to perform a three-dimensional reciprocal space map, when the alignment of the tilt occurs naturally. The corresponding scattering angle 2ω' can then be converted to an interplanar spacing, d_{hkl} with equation 2.8 after account has been made of the refractive index. The refractive index can be determined by simulation (comparing the Bragg angle and the actual peak position to account for absorption effects) or through this formula:

$$\Delta\{2\theta\} = (n-1)\{\cot(\theta - \varphi) + \cot(\theta + \varphi) + \tan\theta\} \qquad 4.11.$$

where n is the refractive index given by equation 2.114. θ is the Bragg angle and φ is the angle between the scattering plane and the surface plane. Since the angle 2ω' = 2θ, can be measured to very high accuracy with the multiple crystal diffractometer the lattice parameter can be determined quite easily to the part per million level for high Bragg angle reflections. This is sufficient to detect deviations from cubic space groups and detect surface damage, Fewster and Andrew (1998).

The advantage of this method is that it can be used to place the reciprocal space maps on an absolute scale and determine the lattice parameters of mosaic blocks.

4.3.5. *Measuring the curvature of crystalline and non-crystalline substrates:*

The curvature of a wafer can be important to establish the state of strain and hence the stress created by the deposition of a thin film. The principle of the method is to establish how the incident angle must be varied to maintain the scattering condition from different regions of the sample. The difference in incident beam angle from two different regions illuminated by X-rays will give the radius of curvature according to

$$R = \frac{x_1 - x_2}{\omega_1 - \omega_2} \qquad 4.12$$

The angular separation can be determined to very high precision and if the separation between the two measurement points is reasonable the accuracy of this measurement can be very high. There are a few important points to consider in this measurement. If the sample is translated then the movement has to be sufficiently good to prevent any twisting that will cause a rotation in ω. This can be the major cause of error. The positions on the sample is best defined with a knife-edge, this then selects the position and size of the region depending on how close the knife-edge is to the sample, figure 4.18. Since we are only concerned with the change in the incident beam angle the most suitable configuration will use the 2-crystal 4-reflection monochromator to create a parallel beam.

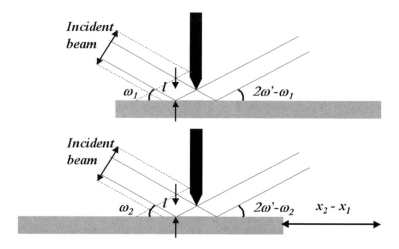

Figure 4.18. A method of determining the radius of curvature by defining the sample position with a knife-edge and translating the sample.

For nearly perfect crystals the angular shift of two peak positions measured at a relatively high scattering angle reflection will yield a reliable result. The sampled region should be as small as possible to reduce peak broadening from the curvature and more precisely define the x value. For the most precise determinations the arrangement given in figure 4.19 should be used. The principle of this method is to translate the knife-edge instead of the sample to overcome the uncertainties in the sample translation. This does rely on a large parallel source which can restrict the separation, $x_1 - x_2$. The choice of reflection and the range of $x_1 - x_2$ may limit the accuracy because of the finite beam size. Clearly a lower order reflection will increase $x_1 - x_2$.

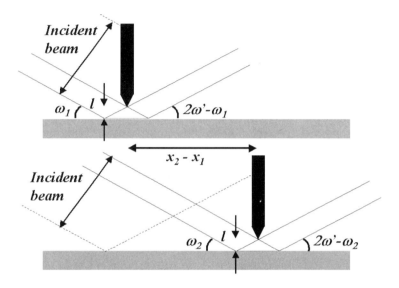

Figure 4.19. The method of determining the curvature by translating the knife-edge to overcome any problems an imperfect sample translation stage.

If the substrate is not crystalline or poorly crystalline then the radius of curvature can be obtained by using the principle described in section 4.3.1.2.3. A relatively large parallel beam from the 2-crystal 4-reflection monochromator incident on the sample is used in conjunction with a knife-edge to define the area on the sample as in figure 4.19. A suitable scattering angle, $2\omega' = 1.4^0$, is chosen and the incident beam angle ω_1 that gives the maximum intensity is noted. Translating the knife-edge across the sample to the second position and determining ω_2 will then give the radius of curvature through equation 4.12. The absolute position of $2\omega'$ is maintained constant throughout. Clearly to obtain a reasonably well-defined region on the sample the knife-edge has to be very close to the sample, this will also limit the broadening of the peak from the curvature.

4.4. Analysis of nearly perfect semiconductor multi-layer structures

The definition of a nearly perfect semiconductor layer was covered in Chapter 1, Table 1.2. If the layer is perfect then we would only be concerned with composition as a function of depth and the surface orientation. Orientation has been covered in the last section. For nearly perfect or real samples, we have defects and possibly small tilts and we shall consider these aspects as well in this section. The analysis of periodic multi-layers will be discussed in section 4.4.4, since the approach and assumptions are different.

We shall consider firstly the measurement of composition and then thickness building from simple assumptions and how these give rise to errors. The most complete approach is through simulation to extract all this information and this is the desired approach, however some of the initial discussion is very useful for determining approximate working values for modelling the scattering. The most sensitive methods rely on the relationship between strain and composition and this will be the main concentration here, which is scattering from Bragg reflections. Reflectometry as a route to composition measurement will be considered for more imperfect structures and periodic structures.

4.4.1. The first assumption and very approximate method in determining composition:

Any material phase will have certain overall characteristics associated with the periodic repeat unit, i.e. lattice parameters $a, b, c, \alpha, \beta, \gamma$. For example AlAs, GaAs, InP, InAs, GaSb, etc., all have the same arrangement of atoms in their unit cells yet the atomic radii differ and consequently the length a $(= b = c)$ is different for each phase. The angles α, β, γ are all 90^0. These differences in lattice parameter through Bragg's equation give rise to different scattering angles. For nearly perfect layers this difference in lattice parameter is accommodated by elastic distortion, see figure 1.6, and section 1.6. The strain in the layer can be expressed as a function of $\Delta\theta$ from the Bragg equation. If the layer is strained to fit to the substrate then the crystal planes originally inclined to the surface plane (by angle φ) will be rotated with respect to the equivalent substrate planes (by angle $\Delta\varphi = (\varphi_L - \varphi_S)$; the subscripts refer to the layer and substrate). Hence

$$\left(\frac{\Delta d}{d}\right) = \frac{\sin(\varphi + \Delta\varphi)}{\sin\varphi} - 1 = -\frac{\sin\theta - \sin(\theta + \Delta\theta)}{\sin(\theta + \Delta\theta)}$$

d is the inter-planar spacing of the substrate and $\Delta d = (d_L - d_S)$ is the difference in the inter-planar spacing for the reflection measured. The difference in the incident angle from the layer and substrate scattering planes is the sum of these contributions; $\Delta\omega = \Delta\theta + \Delta\varphi$. If we assume that $\Delta\varphi$ and $\Delta\theta$ are small then from the last equation $\Delta\varphi \sim \tan\varphi \cot\theta \, \Delta\theta$ and we can write

$$\left(\frac{\Delta d}{d}\right) = \frac{\Delta\omega}{\tan\theta + \tan\varphi} \qquad 4.13$$

Again considering the geometry of the rotation of the crystal planes for a perfectly strained layer on a substrate we write

$$\left(\frac{\Delta d}{d}\right)_\perp = \frac{\tan(\varphi + \Delta\varphi)}{\tan\varphi} - 1$$

The perpendicular sign refers to the equivalent value normal to the surface. Bringing these terms together we have

$$\left(\frac{\Delta d}{d}\right)_\perp = \left(\frac{\Delta d}{d}\right) \frac{\left\{\dfrac{\tan(\varphi + \Delta\varphi)}{\tan\varphi} - 1\right\}}{\left\{\dfrac{\sin(\varphi + \Delta\varphi)}{\sin\varphi} - 1\right\}} \sim \left(\frac{\Delta d}{d}\right) \frac{1}{\cos^2\varphi} \qquad 4.14$$

An additional influence on the scattering plane rotation can result from a layer grown on a terraced surface, e.g. on a substrate surface that is not parallel to the (001) planes. The layer interplanar spacings will be constrained to the step surface (e.g. the (001) plane) and also to the step edge such that the expected tetragonal distortion is reached over the expected step length $<L>$. The assumption is that no relaxation has occurred. If the step edge height is h then the step length will be $<L> =$

$h/\tan\varphi$, where φ is the macroscopic surface to crystal plane angle. The rotation due to this stepped surface is given by

$$\Delta\varphi' = \tan^{-1}\left(\frac{\Delta h}{<L>}\right) = \tan^{-1}\left(\left(\frac{\Delta h}{h}\right)\frac{h}{<L>}\right) = \tan^{-1}\left(\left(\frac{\Delta d}{d}\right)\tan\varphi\right)$$

where $\Delta d/d$ is the strain along the principal crystallographic direction in the layer with respect to the substrate and is equivalent to $\Delta h/h$, Nagai (1974), Auvray et al (1989). To establish an order of magnitude of this effect an AlAs layer on a GaAs (001) substrate will give an additional angular separation of 2.4" arc for a 0.25^0 substrate off-cut, 20"arc for 2^0 and 30"arc for 3^0, etc. Clearly as the strain in the layer increases or the off-cut increases the assumptions that no strain relaxation from dislocations occurs becomes invalid and the whole modelling process becomes more complex, see section 4.6.

The perpendicular mismatch can be related to the mismatch parallel to the interface, i.e. that of the substrate to the free-standing value for the layer, section 1.6. From equations 1.7 and rearranging we can state

$$\varepsilon_{//} = \frac{1-v}{1+v}(\varepsilon_{//} - \varepsilon_{\perp}) \qquad 4.15$$

Taking our special case of a cubic (001) orientated sample and figure 1.6, we can write our strains as

$$\varepsilon_{\perp} = \frac{c_L - a_L}{a_L}$$

$$\varepsilon_{//} = \frac{a_S - a_L}{a_L}$$

Substituting into equation 4.15 and multiplying both sides by (a_L / a_S) we have

$$\left(\frac{\Delta d}{d}\right)_{//} = -\frac{1-v}{1+v}\left(\frac{\Delta d}{d}\right)_{\perp} \qquad 4.16$$

where v is a Poisson's ratio for a particular direction. We could equally well define some distortion coefficient given by $(1-v)/(1+v)$, Hornstra and Bartels (1978). This is an appropriate expression for isotropic strain in the plane of the interface when we can relate the strain of the layer to that of the substrate.

For cubic space groups the relaxed strain parallel to the interface is equivalent to the strain in the relaxed cubic lattice parameter for the layer, hence

$$\left(\frac{\Delta d}{d}\right)_{//} = -\left(\frac{\Delta d}{d}\right)_{Relaxed} = -\left(\frac{\Delta a}{a}\right)_{Relaxed}$$

Hence from our X-ray scattering experiment we can determine $\Delta\omega$ and the perpendicular strain using equation 4.13 and 4.14. The parallel mismatch can be determined from knowledge of the elastic constants. The parallel mismatch relates the atomic layer separation of the free-standing material to the strained value that matches to the substrate.

The X-ray experiment in its is simplest form can be undertaken with a double crystal diffractometer (section 3.8.1). The scattering angle of the collimator crystal should match that of the sample scattering angle to limit the wavelength dispersion. The large detector window has the advantage of collecting the intensity from both layer and substrate when the scattering peaks are well separated (e.g. this becomes more critical when the scattering planes are inclined to the surface plane and small angles of incidence). For scattering planes parallel to the surface plane then a double slit diffractometer can be used, section 3.7.1. However the wavelength dispersion can add confusion (effectively figure 3.6 is superimposed on each peak). The peak separation increases with scattering angle and therefore the sensitivity is improved at higher scattering angles.

Clearly knowledge of the parallel and perpendicular mismatch and the elastic constants allows us to determine the value of $(\Delta d/d)_{Relax}$ and this can be compared with the value expected for a given composition.

For most semiconductor systems we can assume that the composition follows Vegard's rule (i.e. the lattice parameter of a mixed phase, which can exist across the full solid solubility range, is given by the sum of the lattice parameters of all the phases present each multiplied by their proportions). The elastic parameters should be scaled in a similar manner. Hence for a ternary alloy $A_xB_{1-x}C$ with a known inter-planar spacing d_{ABC} and constituent binary compounds with interplanar spacings d_{AB} and d_{BC} we can write

$$x = \frac{d_{DE}\left[\left(\frac{\Delta d}{d}\right)_{Relax} + 1\right] - d_{BC}}{d_{AC} - d_{BC}} \qquad 4.17$$

where $\Delta d = (d_{ABC} - d_{DE})$, the difference in the inter-planar spacing between the layer and the substrate and $d = d_{DE}$ the inter-planar spacing within the substrate. For a sample that is cubic in its natural free standing state then $a \propto d$ and therefore a can be substituted for d throughout these equations.

The assumptions are that the relative orientation between the substrate and the layer is zero, that the inter-planar spacings and elastic parameters follow Vegard's rule and that the scattering can be simply interpreted in this way, i.e not influenced by dynamical diffractions effects. Also since this is a relative measurement the inter-planar spacing of the substrate (the internal reference) is assumed to be a known value.

In general the deviations of the angular spacing from linearity (this brings together lattice parameters and elastic parameters) is very small and very difficult to measure. Most of the problems are associated with finding a better method of measuring the composition for comparison; some of these methods have been reviewed in Fewster (1993). The Si – Ge alloy system has been studied very closely by Dismukes et al (1964) and can be represented more precisely by a polynomial relationship, there is perhaps a stronger justification for using this relationship since the deviation is nearly 3% in the worst case. The expression is given by

$$a_{Ge_xSi_{1-x}} = xa_{Ge} + (1-x)a_{Si} + 0.007\{(2x-1)^2 - 1\} \qquad 4.18$$

If we assume Vegard's rule then the composition will be underestimated. Vegard's rule is applied to the elastic parameters and is sufficient within the accuracies that these are known.

The lattice parameter assumption concerning the substrate is very important, but as described in section 4.3.4 we can determine this independently and thus remove this uncertainty. Another very important assumption concerns the scattering mechanism itself and this will be discussed in the next section.

The discussion so far has concentrated on the single variable that is an alloy that exists between two binaries. However more complex phase mixtures are of interest and these either require additional information, e.g. photoluminescence determination of the energy gap, or additional X-ray scattering experiments. This whole analysis until now has concentrated on the peak positions with the simple Bragg equation transform from scattering to real space; however the scattering strength is not considered. This deficiency again points us to the power of simulation methods that include all aspects.

4.4.2. The determination of thickness:

Interference fringes observed in the scattering pattern, due to the different optical paths of the X-rays, are related to the thickness of the layers. From section 2.7.2 we derived the Scherrer equation 2.94 that relates the width of the diffraction peak to the thickness of the independently scattering region. From section 2.8 we indicated that the reflectometry profile also contains fringing that relates to the thickness of the layers. These appear attractive routes to determining the thickness, however we have to be cautious of these approaches as will be discussed below. Despite this, simulation methods are very reliable.

4.4.2.1. Determining the thickness from the fringes close to main scattering peaks:

From figure 4.20a we can see that the profile from a relatively simple structure is basically composed of a substrate peak, layer peak and thickness fringes. We can determine the thickness fringe peak positions n_1, n_2, n_3,... etc., and determine a characteristic length scale, L, from the Bragg equation which gives the simplest transform from scattering to real space. Now we write for scattering from planes parallel to the surface:

$$2L\sin\omega_1 = n_1\lambda$$
$$2L\sin\omega_2 = n_2\lambda$$

Hence

$$L = \frac{(n_1 - n_2)\lambda}{2(\sin\omega_1 - \sin\omega_2)} \sim \frac{(n_1 - n_2)\lambda}{2\Delta\omega\cos\omega_1} \qquad 4.19$$

Where $\omega_1, \omega_2,...$ correspond to the angular positions of the peaks and $\Delta\omega = \omega_1 - \omega_2$, and ω is some average of the two values. This equation is valid for scattering from planes parallel to the surface.

For the more general case we can derive an expression from the reciprocal lattice construction of figure 3.3:

$$L = \frac{(n_1 - n_2)\lambda}{\{\sin\omega_1 - \sin\omega_2 + \sin(2\omega_1' - \omega_1) - \sin(2\omega_2' - \omega_2)\}}$$
$$\sim \frac{(n_1 - n_2)\lambda}{\{\cos\omega\Delta\omega + \cos(2\omega' - \omega)(\Delta\{2\omega'\} - \Delta\omega)\}} \qquad 4.20$$

Table 4.1. Errors in the assumption of equation 4.19 due to diffraction effects.

Simulated thickness (μm)	Equation 4.19. derived thickness (μm)	Error (%)
2.0000	1.9597	-2.02
1.0000	0.9798	-2.02
0.5000	0.4899	-2.02
0.2500	0.2398	-4.08

If we take a series of layers of different thicknesses, simulate the profile and derive the thickness with equation 4.19 we can compare the peak separation method directly, Fewster (1993). The error can be as much as 4%, and varies with thickness, Table 4.1. Remember of course that this is the simplest case and therefore making assumptions based on more complicated multi-layers does become problematic and more unreliable.

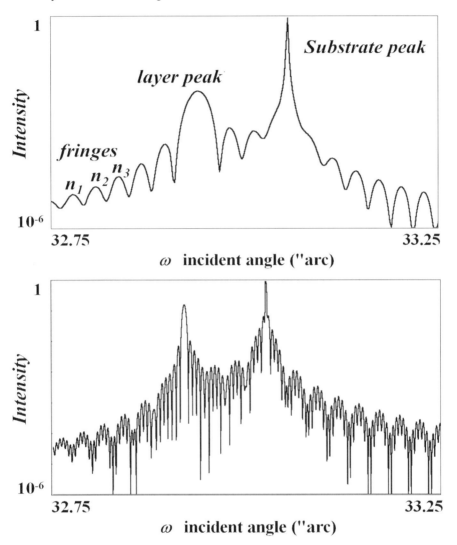

Figure 4.20. (a) The scattering from a single layer on a substrate illustrating the peak broadening and fringing from the limited layer thickness. (b) The scattering from a two layer structure indicating the two modulations. Both simulations indicate the degree of complexity in the scattering profile leading to peak-pulling effects and inaccuracies in making direct determination of structural parameters.

As with the composition determination from peak separation, section 4.4.3.1, this method will only give an approximate value, a rough guide or an input to a simulation model.

4.4.2.2. Determining the thickness from the fringes in the reflectometry profile:

Fringes are very evident in a reflectometry profile, figure 2.26 and as pointed out by Keissing (1931) these should relate directly to the thickness of the layer causing them. As we can see from Table 4.1 the errors due to dynamical scattering effects can be very large and therefore it is not surprising that these errors also appear in the direct determination of reflectometry scans. By way of an example consider a 30nm layer on a substrate and compare the thickness measured with equation 4.19 with that obtained from simulation and see the very large differences, Table 4.2.

Table 4.2. The errors in the direct interpretation of reflectometry profiles for thickness based on equation 4.19 for a sample with a 30nm layer of GaAs on Si.

Mean angle of measurement (degrees)	Thickness determined directly (μm)	Error (%)	Thickness determined with n (μm)	Error (%)
0.4285	0.0442	47.3	0.0291	3.0
0.6626	0.0338	12.7	0.0294	2.0
0.9332	0.0317	5.7	0.0299	0.3
1.2151	0.0309	3.0	0.0302	0.7

The errors decrease rapidly with increasing scattering angle and therefore any measurements based on this direct interpretation should be done at the highest angles possible. The cause of this is related mainly related to the refractive index effects that must now be included. If we consider Bragg's equation 2.9 to determine the path length associated with the peak of the m th order fringe and consider the plane spacing as the thickness of the layer, t, then

$$2t \sin \theta_l = m \frac{\lambda}{n}$$

θ_l and n are the incident angle for the peak of this fringe and refractive index inside the layer. The incident angle on the layer surface, θ, is related to θ_l through Snell's rule, equation 2.121 and the refractive index can be related to the critical angle, θ_c, equation 2.132

$$n = \frac{\cos\theta}{\cos\theta_l} = \cos\theta_c$$

Combining these equations and using the relationship $\sin^2\theta = 1 - \cos^2\theta$

$$m\frac{\lambda}{2t} = n\sin\theta_l = \cos\theta_c \left\{1 - \frac{\cos^2\theta}{\cos^2\theta_c}\right\}^{1/2} = \left\{\cos^2\theta_c - \cos^2\theta\right\}^{1/2} \approx \left\{\theta^2 - \theta_c^2\right\}^{1/2}$$

If we now measure the angle at the peak of two fringes that are M orders apart, as in equation 4.19, then the layer thickness becomes:

$$t = \frac{M\lambda}{2\left(\left(\cos\theta_c^2 - \cos\theta_2^2\right)^{1/2} - \left(\cos\theta_c^2 - \cos\theta_1^2\right)^{1/2}\right)} \approx \frac{M\lambda}{2\left(\left(\theta_2^2 - \theta_c^2\right)^{1/2} - \left(\theta_1^2 - \theta_c^2\right)^{1/2}\right)}$$

The improvement can be seen when we include this into our estimation of our 30nm layer thickness, table 4.2. The effect is very significant for the reflectometry profile. Again simulation is the best method of extracting information from a reflectometry profile, discussed in section 4.4.3.4.

4.4.3. The simulation of rocking curves to obtain composition and thickness

In the previous section we have simply assumed Bragg's equation for each layer and substrate reflection. We could take into account the refractive index but this is a small correction, especially since we are comparing the difference in the positions of peaks, hence any error is related to the difference in refractive index correction. However as pointed out in Fewster (1987) and Fewster and Curling (1987) the peak position from single layers less than 0.5μm or buried layers less than 2μm do not correspond to the peak position expected from Bragg's equation. These differences can be large, giving errors in the composition up to ~15%. The reason for this is that the build-up of the wave-field in the crystal requires a reasonable sustained periodicity to lock into and for thin layers this has not been established. As with all the arguments expressed so far we cannot isolate aspects in the experimentally determined profile and assign them unambiguously to certain features. All we can state is that some scattering feature is predominately influenced by some layer. Dynamical theory considers the whole process of scattering as wave-fields including all the interactions and therefore this becomes the most exacting way of describing the scattering. To illustrate the effect, consider figure 4.20a and b for single and buried layers.

Both profiles give more peaks than there are layers and result from a mixture of thickness fringes (interference from the lower and upper interfaces of a layer) and interaction between all these contributions. The scattering process is therefore quite complex and the simulation and comparison is the most reliable approach. At this stage we are assuming the material to be nearly perfect. Since we are dividing the structure into layers of constant composition and hence strain, we can simply use the same approach for any strain distribution, for example ion-implanted materials, (Sevidori, 1996: Klappe and Fewster, 1993). The method is very general.

To indicate the level of information obtainable from careful and rapid data collection and analysis consider the data from a SiGe heterojunction bipolar transistor, HBT, described below.

4.4.3.1. Example of an analysis of a nearly perfect structure

We wish to collect and analyse the structural details as quickly and accurately as possible and this example should give an indication of what can be done. This SiGe HBT is a two layer structure (a thin (0.1μm) $Ge_{0.15}Si_{0.85}$ alloy layer wedged between a Si substrate and a Si cap layer (~0.7μm). The simplest and easiest experiment is the 004 rocking-curve since the structure is grown on a (001) surface. The instrument was configured with a 2-crystal 4-reflection monochromator with Ge crystals and symmetrical 220 reflections and an open detector. The intensity at the detector at the zero position without the sample is about 4 million counts per second. The procedure is as follows:

1. The sample is mounted on the support plate.
2. An absorbing (nickel, copper) foil is placed between the monochromator and X-ray tube. The detector is at $2\omega' = 0$.
3. The sample is driven into the X-ray beam to reduce the count rate by half. Since the incident beam may not be passing parallel to the surface, the sample is scanned in ω and held at the position of maximum intensity. The sample is driven again to reduce the original intensity by half. The $\omega = 0$ is now set and the sample is in the centre of the circles.
4. The absorbing foil is removed.
5. The detector ($2\omega'$) and sample (ω) axes are driven to the approximate angles for the Si 004 reflection ($\omega \sim 34.5^0$ and $2\omega' \sim 69^0$ for CuKα_1 radiation).
6. The detector is held stationary and the sample scanned very quickly over about 1^0 to find the strong substrate peak reflection. When found the sample axis is fixed at this position.
7. The tilt axis is then scanned over a couple of degrees (χ axis, sections 3.8.1.2, 3.8.3.2) and the mid-cord of the resulting profile defines the aligned tilt angle. A small drive down in the sample axis will recover the intensity if this profile is double peaked.
8. The GeSi layer will have a larger lattice parameter than the substrate Si and therefore the rocking curve will be centred on the low angle side of the substrate. A 5-minute rocking-curve is then performed and the result is as in figure 4.21.

Step 2 to step 4 are not essential but can avoid difficulty in finding the reflection in step 6. The ω angle will have some uncertainty and therefore a larger rapid search scan may be necessary (step 6) and the sample centring will ensure that the scattered beam is closer to the true $2\omega'$ angle for the substrate.

We shall now concentrate on two approaches; the very direct method given above and simulation (manual and automatic).

212 X-RAY SCATTERING FROM SEMICONDUCTORS

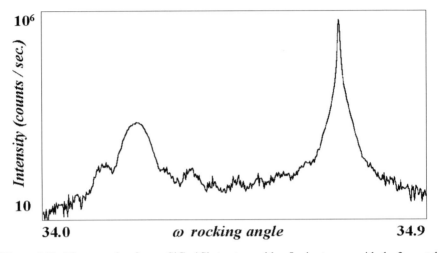

Figure 4.21. The scattering from a SiGe / Si structure with a 5 minute scan with the 2-crystal 4-reflection monochromator and open detector.

4.4.3.2. Direct analysis from peak separation and fringe separations

We can assume that for this relatively simple structure that the strongest and second strongest peaks come from the substrate and GeSi layer respectively. The top Si layer should be hidden under the substrate peak if the layers are all strained with respect to the substrate. The peak separation yields a composition of 18.2% Ge in the layer using equations 4.13 to 4.17. The fringes between the layer and substrate yield a thickness of some layer in the structure of 0.10±0.01μm using equation 4.19. From our knowledge of the structure this will almost certainly be associated with the GeSi layer.

Now we know that neither of these assumptions is precise, also we do not know the thickness of the cap layer, yet this appears somewhere in the profile.

4.4.3.3. Simulation using an iterative adjustment of the model

Initially we can either start with the found values from the direct analysis or take the assumed values from that expected from the growth. If we take the latter approach then the peak position of the layer is a long way from the best-fit value and therefore we should get these aligned first. The fitting of the thickness will now be more precise since the match is to the broadened shape of the layer and the fringe separation. The Si cap layer will have a much more subtle effect on the profile, since its thickness is ~7× larger than the SiGe layer the associated fringing will be oscillating at ~7× the frequency. The quality of the data may not be too good to fix this parameter too precisely. Clearly if the thickness is considerably smaller than expected, 0.3 or 0.4µm for example, then the observed fringes from the SiGe will be modified such that their rise and fall can be slightly different giving the appearance of displacing the fringes. However if the cap layer is considerably thicker than expected then the influence on the pattern is insufficient to detect with this data quality for this particular example.

4.4.3.3.1. Linking parameters to cope with complex multi-layer structures

If there is good control over the growth method then we can simplify the analysis by linking the growth rates of the various phases deposited. This can prove very useful for analysing very complex structures and will also aid automatic approaches to fitting profiles, Fewster (1990). Suppose the growth rate of phase A (e.g. GaAs) and phase B (e.g. AlAs) in layer L are denoted by $R_A(L)$ and $R_B(L)$ and these are consistent throughout growth. Then the timing of the growth for each layer can be directly related to the layer thickness. The "sticking coefficient" of each phase should be known and for many materials this can be considered as unity, should it be less than this but always consistent then this will not influence this approach, just change the effective growth rate. To emulate the growth process precisely we should include transients due to the temperature differences of an open Knudsen cell compared with a closed one in an MBE growth chamber. The high temperature and hence higher growth may occur at the beginning of each layer deposited and this may become significant in the comparison of growth rates for very thin and very thick layers. We therefore have for a mixture of two binary phases

$$t_A = T_L R_A(L) V_1$$
$$t_B = T_L R_B(L) V_2$$

4.21

where t_A and t_B are the thickness contribution of binaries A and B. T_L is the time for the layer to be grown and V_1 and V_2 are variables to be refined (in effect the modification to the expected growth rates). Then the total thickness of the layer is given by

$$t_L = t_A + t_B \qquad 4.22$$

and the composition or proportion that is phase B is given by

$$x_L = \frac{t_B}{t_L} \qquad 4.23$$

Since this is a volume concentration the assumption here is that the two materials have similar lattice parameters. We have now related the thickness to the time that the shutters are open. This timing is far more precise than knowledge of the various growth rates. To include the transients we can assume an exponential form to reach equilibrium:

$$\begin{aligned}R_A(L) &= R_A(L-1) - [R_A(L-1) - {}_0R_A](1 - e^{-A_0 T_L}) \\ R_B(L) &= R_B(L-1) - [R_B(L-1) - {}_0R_B](1 - e^{-B_0 T_L})\end{aligned} \qquad 4.24$$

${}_0R_A$ and ${}_0R_B$ represent the equilibrium growth rates when the shutter is open and A_0 and B_0 are the characteristic cooling rates of the cells and are further parameters to be refined. Of course since the cells cool from being open they must also warm whilst being closed and therefore the growth rate must also be considered to take into account the equivalent equilibrium growth rates, ${}_cR_A$ and ${}_cR_B$ with the shutters closed:

$$\begin{aligned}R_A(L) &= R_A(L-1) - [R_A(L-1) - {}_cR_A](1 - e^{-A_c T_L}) \\ R_B(L) &= R_B(L-1) - [R_B(L-1) - {}_cR_B](1 - e^{-B_c T_L})\end{aligned} \qquad 4.25$$

where A_C and B_C are the characteristic warming rates. These characteristic rates of cooling and warming should be constant and in a well characterised

MBE system known. We therefore have two additional variables to include in any refinement associated with each phase.

For well-controlled growth and large Knudsen cells the effects of transients is less significant. This example does indicate the detail that can be extracted or alternatively an effect that we should be aware of especially in the growth of very short period superlattices. The general idea though of linking growth rates is very useful for rapid evaluation of complex multi-layer systems with mixed phases.

The iterative process can be fast with experience, however this whole process can be greatly aided by automatic fitting of the data.

4.4.3.4. Automatic fitting of the data by simulation

If we want a fairly "push button" method so we work from the basis that the growth model is the starting value. The fitting of the data is a non-linear problem and full of false minima and so is not a simple procedure. Non-linear least squares procedures have been used successfully when a very good estimate of the model is known already, e.g. when all the direct analysis information is included, Fewster (1990). The basic problem is to somehow find a position close enough to the global minimum (the minimum error possible) or have an algorithm that will allow movement out of local minima (false minimum errors) so that the true best fit to the global minimum is found. The global minimum will contain the parameters associated with the true wanted model of the structure.

There are several approaches to finding the global minimum. Simulated annealing algorithms work on the basis of allowing any solution to move out of local minima and the search will progress by refinement of the model until the global minimum is found (or what it thinks is the global minimum). Genetic algorithms also offer possible routes to the solution of scattering profiles, (Dane, Veldhuis, de Boer, Leenaers and Buydens, 1998). This procedure works on a rather random process of trial structures that favours the closest fits to the model and combine parameters to create another set of trial structures. Of course these methods may find the global minimum by chance. If they find local minimum then the problem is overcome by randomly changing some parameters in the hope of moving out of these. The whole fitting process can vary in timing because of the random nature of the algorithm and difficulties in judging the end point, Wormington, Panaccione, Matney and Bowen (1999).

Another approach that we shall discuss here is an algorithm that reduces the presence of local minima and can find the correct solution exceedingly rapidly, Fewster (1993). This approach proved to be very successful for analysing ion-implanted layers, Klappe and Fewster (1993) and works very well for heterostructures, Tye and Fewster (2000). The underlying assumption is that the amount of detail in the data is initially too great and by fitting the main features an approximate solution close to the global minimum can be found. This method is ideally suited to rocking curve simulation. For reflectometry profiles the occurrence of false minima is again clear and this was the main problem that Dane et al (1998) were trying to solve. However an alternative approach based on a different principle is preferred, Fewster and Tye (2001). This is a segmented fitting routine that assumes that the overall structural form is contained in the first part of the reflectometry curve and so this is fitted first. When this fits well, more of the profile is included bringing in more detail. This approach steers the automatic fitting to the global minimum.

However we are mainly interested in the fitting of rocking curves so we shall concentrate of the method of Fewster (1993). Consider our profile of figure 4.21, the main features are the layer and substrate peak. The algorithm works by simulating the profile with a starting model and then heavily smoothing this calculated and experimental profile for comparison, figure 4.22a. Using a steepest descent algorithm the agreement between these two profiles is achieved very quickly, figure 4.22a. The procedure is repeated with this improved model but with less smoothing. A solution is found in a few seconds on a personal computer, figure 4.22b. The agreement is based on a logarithmic comparison of intensities and because of the smoothing the influence of noise, data quality and material quality (peak broadening) have less effect than direct comparisons. The background can be added and fixed or refined.

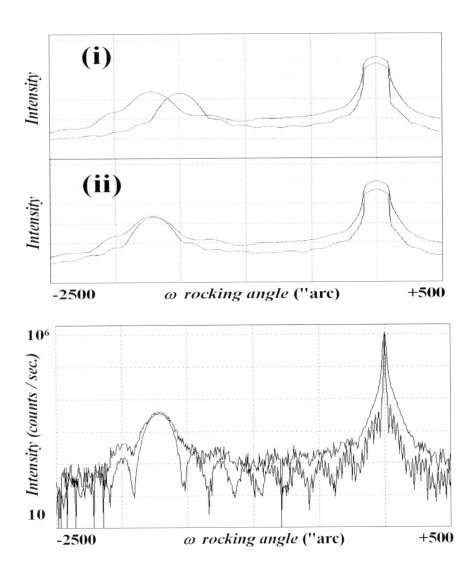

Figure 4.22. (a) The principle of the fitting algorithm; (i) the simulation of the expected structure compared with the measured profile after both have been smoothed. (ii) because of the removal of noise and the removal of local minima the first stage of the refinement is very rapid. (b) The final fitted profile after the smoothing has been removed.

The final fitted parameters are 18.51(2)% Ge in the GeSi layer that is 0.101(1) μm thick, capped with a Si layer of 0.70(5)μm. The composition is higher than in the direct analysis and this is the peak pulling effect discussed at the beginning of section 4.4.3. The thickness determined by simulation for this SiGe layer is in close agreement with the direct method. A value is given for the cap thickness, although as discussed it is not well determined in this experiment.

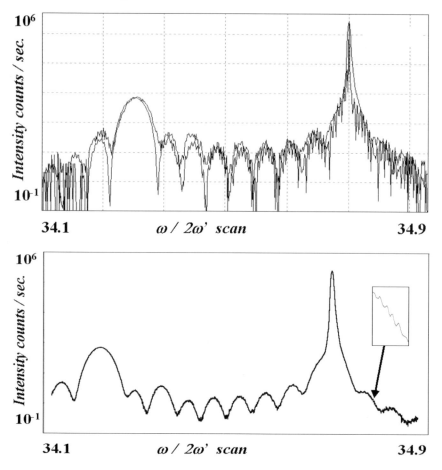

Figure 4.23. (a) A repeat of the procedure illustrated in figure 4.22b but with a multiple-crystal profile taken in 5 minutes. (b) The same profile but taken over 1.5 h, illustrating the fine fringing that relates to the thick cap layer.

Let us consider the profile more closely. The layer and fringing appear to fit very well, but close to the substrate peak the fit is poor. We could include some additional diffuse scattering into the model (a $(1/\Delta\omega)^2$ dependence profile with a certain height and width, section 2.7.4), although in this case this makes little difference. Sometimes this can help to achieve a more rapid and exact fit in some structures although it is not relevant to a greater qualitative physical understanding. We shall now repeat the experiment except with an analyser, taking the same length of time for the measurement, i.e. 5 minutes.

4.4.3.5. Data collection with the 2-crystal 4-reflection monochromator and 3-reflection analyser

The experiment set-up is as in figure 3.27, initially without the X-ray mirror. The experiment was performed following this procedure:
1. The sample is mounted on the support plate.
2. *An absorbing foil is placed between the monochromator and X-ray tube. The detector is at $2\omega' = 0$.*
3. *The sample is driven into the X-ray beam to reduce the count rate by half. Since the incident beam may not be passing parallel to the surface, the sample is scanned in ω and held at the position of maximum intensity. The sample is driven again to reduce the original intensity by half. The $\omega = 0$ is now set and the sample is in the centre of the circles.*
4. The absorbing foil is removed.
5. The detector ($2\omega'$) and sample (ω) axes are driven to the approximate angles for the Si 004 reflection ($\omega \sim 34.5^0$ and $2\omega' \sim 69^0$ for CuKα_1 radiation).
6. The detector is held stationary and the sample scanned very quickly over about 1^0 to find the strong substrate peak reflection. When found the sample axis is fixed at this position.
7. The tilt axis is then scanned over a couple of degrees (χ axis, sections 3.8.1.2, 3.8.3.2) and the mid-cord of the resulting profile defines the aligned tilt angle. A small drive down in the sample axis will recover the intensity if this profile is double peaked.
8. An ω scan is performed to precisely locate the peak of the substrate reflection.

220   X-RAY SCATTERING FROM SEMICONDUCTORS

9. The analyser / detector assembly is then substituted for the open detector.
10. The $2\omega'$ (analyser / detector) axis is then scanned to detect the scattered beam. This will be a very sharp peak and so a small step size is needed for the final location of the peak position.
11. With the two axes set to detect the scattered X-rays they can be driven together. Since we are studying the 004 reflection from a 001 sample the ratio can be maintained at 1 : 2 for $\omega : 2\omega'$. An $\omega/2\omega'$ scan is carried out over the same range of ω as in the previous experiment and we achieve a perfectly useable profile in 5 minutes. However the fringing is rather weak and is greatly improved by including the mirror. The data are collected again in 5 minutes and shown in figure 4.23a.

The procedural steps 1 to 7 are identical to those needed to collect a rocking curve. Again steps 2 to 4 are not essential.

Clearly the profile is superior and now we can accurately fit the profile close to the substrate. The results become ($x = 18.58(18)\%$, $t_{GeSi} = 0.098(1)\mu m$, $t_{cap} = 0.717(3)\mu m$). The simulation does not include background or diffuse scattering and assumes that the sample is perfect. The top cap layer, as mentioned previously is rather thick and leads to very high frequency fringing that can be observed with a more careful scan of 1.5 h (small step size and longer counting times) to yield the fringing and thickness, figure 4.23b. The advantage of this experimental configuration is now clear. However it may be important to know the reason for the difference in the profiles of figure 4.23 and figure 4.21. This all becomes clear when we perform a reciprocal space map.

4.4.3.6. Reciprocal space map to analyse the imperfections in samples

Having set up the previous experiment with the monochromator and analyser (i.e. procedure 1 to 10) we can now carry out a reciprocal space map. Rather than conducting a single $\omega/2\omega'$ scan we should perform several of these with slightly different settings. Continuing on from the action number 10 above, and since we know that the intensity close to the substrate is broader than it should be a suitable next step is:

11. Scan in ω over a reasonable range over the substrate peak to see how far the intensity spreads.
12. Use this angular range as a guide to the spread in the width (ω) of the scattering to be captured by reciprocal space mapping.
13. The reciprocal space map is then collected over the angular range given by the ω / 2ω' scan of the previous experiment and the ω spread of step 11.
14. The resulting reciprocal space map can be collected either at a similar or shorter count-time to the previous experiment depending on the detail of the analysis. For this purpose it was collected with 1s per point. There were 1000 steps in ω / 2ω' and after each scan the ω was offset by 0.0015. This was repeated 60 times to give a respectable reciprocal space map in 17h, figure 4.24.

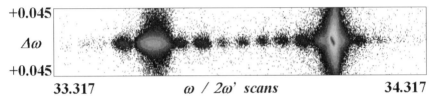

Figure 4.24. The reciprocal space map of the SiGe / Si sample, indicating the significant diffuse scattering close to the substrate and layer reflections.

The reciprocal space map shows significant diffuse intensity around the substrate peak and the layer peak. We can take a guess at the cause of this additional scattering or we can build on our picture by placing the instrument probe (i.e. setting the ω and 2ω' angles) to be on these "wings." If we now place a topographic emulsion after the analyser we will see which parts of the sample contribute to this scattering, figure 4.25. What we observe is a crosshatch of lines that lie along the <110> type directions. We can assume from this that defects are forming at the GeSi / Si substrate interface and these are 60^0 dislocation that are partially relieving the internal stresses in the sample. The image width of these lines in the scattering plane is about 7μm and in the axial direction 100μm. These lines are predominately the strain-fields from single dislocations. We can understand the whole scattering shape to arise from the local stretching of the Si substrate lattice parallel to the interface plane; this reduces the lattice parameter normal to the interface through the Poisson effect. Consequently the scattering angle is increased for this region and the tips of these "wings" are at higher angles than the main peak. The spread in the ω direction arises from the curvature of the scattering planes close to the dislocation. The very high strain sensitivity of this experimental configuration means that the distortion is detected a long way from the dislocation core (i.e. many microns). This scattering is only that which comes from between the perfect region and the outer limits of the dislocation strain-field and explains the very small angular spread of the scattering. The analysis close to the layer peak can be carried out in a similar way.

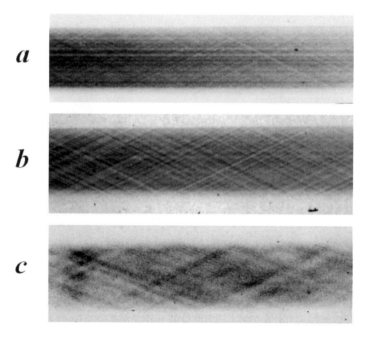

Figure 4.25. A series of topographs (a) and (b) with the emulsion placed immediately after the sample at the layer and substrate peak respectively (the later is a 5 minute exposure L4 emulsion). The topograph (c) is obtained on the diffuse scattering region close to the substrate with the emulsion placed after the analyser (2h exposure). Note the cross-hatch of lines parallel to the [110] directions typical of relaxing layers.

Since the individual dislocations are clearly observed we can count the number per unit length and determine the degree of relaxation. We have to make an assumption on the type of dislocations that exists in the sample and in Si these are almost entirely 60^0 dislocation with a Burgers vector, b, 45^0

to the interface plane. Each dislocation will therefore take out a line of atoms equivalent to $b_{//} = d_{[110]}/2 = a/\{2\sqrt{2}\}$. In this case we can measure m (~12) dislocations in a length M (~230μm) in two orthogonal directions, giving a dislocation density of $\rho_{disl} \equiv \dfrac{2m}{M^2} = 4.54 \times 10^4 \mathrm{cm}^{-2}$. The mismatch and dislocation density in the plane of the interface is given by

$$\left(\frac{\Delta d}{d}\right)_{//} = \frac{mb_{//}}{M} = \frac{m\dfrac{a}{2\sqrt{2}}}{M}$$

$$\rho_{disl} = \frac{2\left(\dfrac{\Delta d}{d}\right)_x \left(\dfrac{\Delta d}{d}\right)_y}{\left|b_{//}\right|^2}$$

4.26

In this case the mismatch is ~2×10^{-5}. The actual mismatch of free-standing $Ge_{0.19}Si_{0.81}$ and Si is ~ 7.93×10^{-3}, giving the relaxation % as

$$R = \frac{2 \times 10^{-5}}{7.93 \times 10^{-3}} \sim 0.26\%$$

4.27

This is the only convenient way to measure very small relaxations, since the influence of the scattering pattern as we will see later is inadequate to change the peak separations. For higher relaxation a peak shift is detectable and the method of counting dislocation lines fails because the individual lines overlap and cannot be resolved.

Another feature of the reciprocal space map is that the layer and substrate peaks do not lie on the same $\omega / 2\omega$' scan. This is because the layer is tilted with respect to the substrate. The effect is very small, only 7.2"arc difference. However we ought to be aware of the difference that this can cause in our analysis. The tilt angle adds or subtracts directly from our peak separation in the rocking curve with no analyser. In our previous rocking curve analysis we had not taken this into account and this we shall do now.

4.4.3.7. Taking account of tilts in rocking curve analyses

Returning to our GeSi sample we can determine the influence of tilt on our composition determination. In this particular example the tilt is very small; from the reciprocal space map we can see that it amounts to 7.2"arc. From figure 4.21 we can see that a layer tilted with respect to the substrate by $\Delta\varphi$ will decrease or increase the peak separation $\Delta\omega$ depending whether the tilt is towards or away from the incident beam direction. Therefore the average angular separation from two rocking curves of the same set of scattering planes with opposite beam paths will have the correct value based on a simple analysis, section 4.4.1.1. Similarly the difference in angular separation $\Delta\omega$ for the two rocking-curves will correspond to $2\Delta\varphi$. We therefore have two choices: include $\Delta\varphi$ into the model of the structure or model both rocking curves and take some average. Clearly the better way would be to include the tilt into the model of the structure since all the parameters will be influenced (composition and thickness).

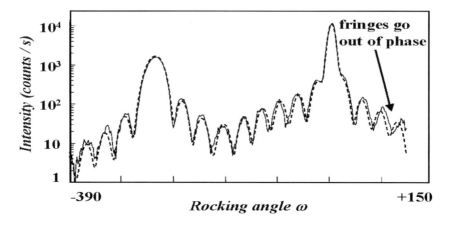

Figure 4.26. The measured and simulated profile indicating some of the problems in achieving an exact fit.

In the GeSi example considered above the difference in the two rocking-curves is barely discernible, because we are observing a 7"arc tilt in a separation of 1850"arc between the layer and substrate. However when the tilt is a significant proportion of the difference we have to take care in the interpretation. The most exacting measurement is achieved with multi-crystal optics, figure 4.23 unless the tilt is too large and one simple ω/2ω' scan will not capture both the layer and substrate peaks. This will be covered in section 4.4.3.8.

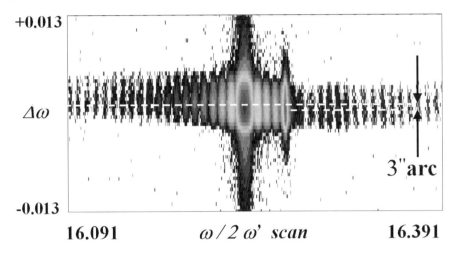

Figure 4.27. A reciprocal space map indicating the subtle kinks in the dynamical streak that leads to the small displacements in the fringing evident in figure 4.26.

Before we leave the problem of analysing tilts in nearly perfect structures we should consider a fairly common problem where a perfect fit can be difficult to achieve. Consider the rocking curve of an AlAs / GaAs superlattice on a GaAs substrate, figure 4.26. The simulated scattering pattern looks almost a perfect fit, however the fringes appear to go out of phase outside the region between the substrate and layer peaks. We also find that the problem exists when we use multi-crystal optics. The reciprocal space map, figure 4.27 reveals a rather strange phenomenon, the fringes appear to have their maxima aligned along the surface normal direction outside the region of the substrate and layer peaks, whereas they lie along a line between the two peaks. The layer and substrate have a small relative tilt. This phenomenon can be simulated when we use a very exact model taking into account the full instrument function response and coherence of all the scattering contributions in the reciprocal space map, Chapter 2. The simulated reciprocal space map illustrating this effect is given in figure 4.28. Except in the most precise analyses we can ignore these effects and take some reassurance that these shifts will reflect a small tilt. The shift of these tilts can be determined geometrically from knowledge of the tilt angle.

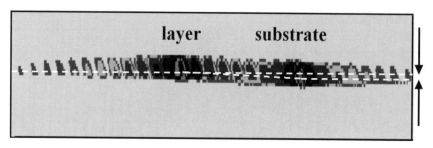

Figure 4.28. The simulation of a reciprocal space map of a structure containing a very small layer misorientation with parallel surface and interfaces.

4.4.3.8. Modelling the extent of the interface disruption in relaxed structures

Using the above analysis we can obtain a good fit to the multi-crystal diffraction profile whereas the rocking-curve shows discrepancies. We have already indicated the source of this effect from topography, i.e. dislocation strain-fields close to the interface of the substrate and GeSi layer. From the strain model and scattering theory described in section 2.5 we can build a better picture of our sample by estimating the extent of the distortions around this interface.

In this particular example we will project all the intensity of the reciprocal space map onto the <001> direction, this then takes out the lattice rotation effects and tilts but includes all the associated diffuse scattering. This will not be identical to the rocking-curve, since the rocking-curve effectively projects the intensity that exists along a line inclined to the direction normal to <001>. However we should be able to account for all the intensity in this profile in terms of thickness, compositions and interfacial strains. A full simulation of the reciprocal space map will give the plane rotations and tilts, etc.

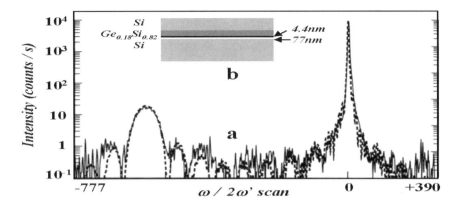

Figure 4.29. The profile obtained by projecting the reciprocal space map to include all the diffuse scattering. The fit to the full profile, including the diffuse scattering gives the full structure including the distortion at the interfaces.

The projected profile is given in figure 4.29. The initial composition and thickness values are taken from the above analysis. Clearly if we are extending the interfacial disruption then this will influence the actual thicknesses. The best-fit profile given in figure 4.29 was obtained with the depth of interfacial disruption into the layer and substrate as 4.4nm and 77nm. The strain has been assumed to take on an exponential form and this disruption depth represents the full extent of this gradation in strain. The thickness and compositions were modified slightly to give a picture of the sample given in figure 4.29b. The shape of the profile close to the interface can therefore be understood in terms of the distortion associated with dislocations close to the interface.

4.4.3.9. Detailed analysis to reveal alloy segregation and the full structure of a multi-layer

The Silicon-Germanium structure described above has a relatively simple form, although the complexity and detail can be quite extensive. We could also see that the analysis of the main structural features could be obtained rather readily. However this next example is of a quantum well laser. The performance was far from expected and this had to be related in some way to the growth of the structure. Initial fast analyses could not reveal the

228   X-RAY SCATTERING FROM SEMICONDUCTORS

problem and therefore very high quality data were required. For this a limited area reciprocal space map was measured close to the 004 reflection and projected onto the <001> direction using the 2-crystal 4-reflection Ge monochromator (symmetrical 220 setting and CuKα_1 radiation) and the 3-reflection analyser crystal. The profile is given in figure 4.30a. From inspection we can point to the "substrate" peak, the AlGaAs "cladding" peak and the very weak "InGaAs quantum well layer" peak. The assignment of peaks relate to their width and scattering strength (broad and weak means thin layer, strong and narrow mean thick layers) and also their relative positions (larger or smaller lattice parameters normal to the surface plane). The quotes refer to the fact that they represent the dominating scattering associated with these peaks.

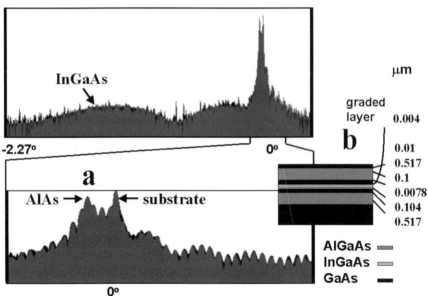

Figure 4.30. The detailed comparison of the measured and simulated profiles gives significant structural information including the segregation of In out of the quantum well.

Approximate values for composition and thickness can be made from the approaches described previously. However an intermediate broad profile could not be modelled successfully. This is where we now bring in the understanding of growth and recognise that *In* can segregate during growth. If this happens the quantum well will be asymmetric, the energy levels will be altered and the emission not what is expected. By introducing a small exponentially falling In concentration towards the surface from the InGaAs quantum well an excellent fit to the whole profile could be obtained. The final model of the structure from the best fit is given in figure 4.30b. This example gives the degree of sensitivity possible with high quality data and persistence in striving for a perfect fit.

4.4.4. Analysis of periodic multi-layer structures:

A periodic multi-layer structure is composed of a sequence of layers that are repeated one or more times to create different physical properties from the individual layers. If the properties of interest see this repeat sequence behaving as a whole then this is a superlattice. Periodic multi-layers create very special properties either as Bragg reflecting stacks for opto-electronic devices or as superlattices that have strong wave function coupling to produce band-folding and new possibilities in quantum effect devices. The periodicity also has a large influence on the X-ray scattering and presents opportunities for extracting information quite directly and for studying interfaces in greater detail.

The structural information required from periodic multi-layers is very similar to that discussed above except that additionally we are interested in variations in periodicity or departures in periodicity and the abruptness and roughness of interfaces. The variation in periodicity and interface quality would influence the reflecting power in a Bragg stack or the quality of the band folding in superlattices. The concentration in this section will centre on these additional parameters.

4.4.4.1. The analysis using direct interpretation of the scattering pattern:

If we simulate the scattering from a periodic structure with dynamical theory we will see, depending on the thicknesses involved, a series of satellite peaks associated with the average composition. We can consider this as a modulation of the composition and the strain about some average value.

From equation 4.19 we also could see that the thickness could be obtained from the separation of the fringes in the scattering profile.

Therefore there should be fringes associated both with multi-layer periodicity and with the overall thickness of the periodic structure. We can determine the period by combining various combinations of satellite positions

$$\Lambda = \frac{(n_i - n_j)\lambda}{2(\sin\omega_i - \sin\omega_j)} \qquad 4.28$$

If the structure is composed of a binary (e.g. GaAs) and a ternary (e.g. $Al_xGa_{1-x}As$) and the thicknesses of the contributing layers are t_{BC} and t_{ABC} then the average composition within the period Λ is given by

$$\bar{x} = x \frac{\frac{\bar{a}}{a_{ABC}} t_{ABC}}{\Lambda} \sim \frac{x t_{ABC}}{\Lambda} \qquad 4.29$$

where

$$\Lambda = t_{BC} + t_{ABC} \qquad 4.30$$

The approximation sign in equation 4.29 arises from the fact that the composition ratio x is related to the unit cell and not to a common unit of length throughout the structure. We have therefore had to relate the number "A" atoms per unit length and translate this to the number of equivalent "A" within the period with average lattice parameter $\bar{a}$. However in the initial analysis we could assume that the unit cell repeat parallel to the growth direction for the two layers are the same. If the periodic structure is composed of two materials of known composition (e.g. two binaries, AlAs and GaAs) then we can consider the "average" structure to be composed of an alloy of the two materials (e.g. $Al_xGa_{1-x}As$). The average composition is hence given by

$$\bar{x} \sim \frac{t_{AB}}{\Lambda} \qquad 4.31$$

where t_{AB} is the layer thickness of the phase AlAs for example.

We can now very simply determine a value of the period, equation 4.20 and the thickness of the individual layers, equations 4.30 and 4.31 for a periodic structure of known compositions in the individual layers if the average composition is determined. The average alloy can be determined approximately from the angular separation of the "average" layer peak and that of the substrate in the rocking-curve, section 4.4.1.

From these equations we can derive some basic information directly from the scattering profile.

4.4.4.2. The analysis using basic kinematical theory:

From the above analysis we can obtain some information, although it does not yield the composition in complex (e.g. three layer repeat) superlattices or interfacial quality. To obtain this information we need to model the intensities. Kinematical theory can prove to be very convenient and fast in extracting this information, Fewster (1986, 1987, and 1998). The basis for this approach is to consider the repeat unit as a unit cell and determine the associated intensity using the kinematical theory approximation.

Firstly we determine the structure factor for the period. The size of the unit cell repeats parallel to the interface can be assumed or measured, but equated to the normal unit cell repeat. However the unit cell perpendicular to the surface is now the repeat period of the superlattice. We can determine this period directly, but of course it may not be equivalent to an integer number of unit cells of a_{BC} and a_{ABC}. In reality it will be composed of several combinations depending on the quality of the superlattice and therefore the period is given by

$$\Lambda = (na_{BC} + ma_{ABC})K_1 + ([n+1]a_{BC} + [m-1]a_{ABC})K_2 + ([n-1]a_{BC} + [m+1]a_{ABC})K_3 \\ + ([n+1]a_{BC} + ma_{ABC})K_4 + (na_{BC} + [m+1]a_{ABC})K_5 + ([n-1]a_{BC} + ma_{ABC})K_6 \\ + (a_{BC} + [m-1]a_{ABC})K_7 + \ldots\ldots$$

4.32

where

$$K_1 + K_2 + K_3 + K_4 + K_5 + K_6 + K_7 + \ldots\ldots = 1$$

4.33

and a_{ABC} and a_{BC} represent the lattice parameters along this direction and m and n are integers. The integers m and n are values that give the closest fit to the period.

$$\Lambda \sim na_{BC} + ma_{ABC} \qquad 4.34$$

We therefore have to calculate the structure factor several times to account for the incommensurability in the structure, depending on the superlattice quality and precision we wish to achieve. If the integers are large then the importance of including a high number of terms is less important because the ratio of the contributing layers are not too seriously effected. The index of reflections associated with each satellite become rather large $\sim (m+n)$ and because the period will almost certainly be incommensurate with the lattice repeat periodicity this index is not exact. It therefore becomes easier to index the satellites as $-2, -1, +1, +2$, etc. with respect to the "average" or "zeroth order satellite" peak for the superlattice.

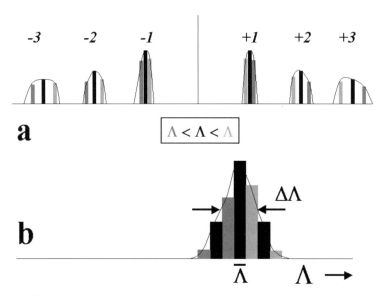

Figure 4.31. (a) A simple explanation of how the distribution of periods in a structure will give rise to the broadening of satellite reflections. The distribution of periods can be approximated to a Gaussian form (b).

We now have to establish the magnitude of the K_j values and this will depend on the variation in period, which is also related to the interfacial roughening, Fullerton, Schuller, Vanderstaeten and Bruynseraede (1992). This can be determined directly, Fewster (1988). From a rather simplistic viewpoint we can imagine a structure composed of several different periods that overall still has some average structure and an "average" scattering peak. Those parts of the structure that have a longer period will have satellites at smaller angles than the average position and those parts that have a shorter period than the average will scatter at larger angles than the average, figure 4.31a. For satellites further away from the "average" peak the angular difference to the "average" satellite position increases and the satellites broaden. If we now differentiate equation 4.28 above for the period to obtain the variation in the period we obtain

$$\Delta\Lambda = \frac{(n_i - n_j)\lambda\Delta\omega_n}{(\cos\omega)\Delta\omega^2} \qquad 4.35$$

where $\Delta\omega$ is the distance between two satellites i and j, and $\Delta\omega_n$ is the angle within one satellite reflection for a difference in period of $\Delta\Lambda$. The satellite order is n (in this notation $n = 0$ is the "average" peak order). This can therefore be measured directly by plotting the satellite order against full-width-at-half-maximum intensity and deconvolving the zeroth order peak width. The zeroth order peak width should be unaffected by period variations and should represent the broadening effect of the data collection method and sample broadening effects. This method assumes that the period variation is random and not systematic. Since this can be measured directly from the width of the satellites this information can be included in the modelling of the satellite intensities. The period variation is assumed to be Gaussian, and the number of unit cells to create a repeat unit is distributed about the mean value, figure 4.31b. The deconvolution of two Gaussian profiles is also rather trivial and straightforward, in the above case we can write

$$\Delta\omega_n = \{\Delta_m\omega_n^2 - \Delta_m\omega_0^2\}^{\frac{1}{2}} \qquad 4.36$$

where $\Delta_m\omega_n$ is the measured peak width and n the satellite order.

The structure factor is calculated using equation 2.22 for all the possible repeat units that straddle the measured period and these are added coherently (maintaining the phase relationships). The intensity is obtained from equation 2.98 taking into account the absorption of the X-rays above the periodic structure. The angular position for the satellite is given by

$$\theta_j = \sin^{-1}\left(\frac{n'\lambda}{2\bar{a}} + \frac{N\lambda}{2\Lambda}\right) \qquad 4.37$$

where N is the satellite order with respect to the "zeroth order satellite" peak and n' is the order of the reflection for the average lattice parameter in the period, $\bar{a}$. Now of course the average lattice parameter $\bar{a}$ is given by

$$\bar{a} = \frac{(na_{BC} + ma_{ABC})}{(n+m)}K_1 + \frac{([n+1]a_{BC} + [m-1]a_{ABC})}{(n+m)}K_2 + \frac{([n-1]a_{BC} + [m+1]a_{ABC})}{(n+m)}K_3$$
$$+ \frac{([n+1]a_{BC} + ma_{ABC})}{(n+m+1)}K_4 + \frac{(na_{BC} + [m+1]a_{ABC})}{(n+m+1)}K_5 + \frac{([n-1]a_{BC} + ma_{ABC})}{(n+m-1)}K_6$$
$$+ \frac{(a_{BC} + [m-1]a_{ABC})}{(n+m-1)}K_7 + \ldots\ldots$$

$$4.38$$

We can therefore very easily determine the intensity associated with each satellite and compare these with the measured intensities. As discussed in section 2.7.2 on the assumptions of the kinematical theory the shape is determined by the thickness and this will smear the intensity that we have just evaluated. The simplest approach is to determine the integrated intensity associated with each satellite and compare this with various model structures. The advantage of this approach is that we can ignore the shape of the peaks and this gives us the possibility to collect the data using many different methods.

Since this approach is not restricted by the contents of the periodic repeat unit we can very easily include variations in composition and several layers in each period. The variation in composition could be interfacial spreading for example and this can take on any shape (Gaussian, linear, exponential or erf) appropriate to how the grading is formed. Also the advantage in just concentrating on the integrated intensities of the satellites is that we can use high intensity low-resolution diffractometers as described in section 3.7.1.

The structure factors of all these various contributions are added as before and then multiplied by its complex conjugate to give the intensity after including the absorption and instrumental aberrations, equation 2.98.

This kinematical theory model of the scattering makes several assumptions. The major assumption is that the structure behaves as some "average" layer with a modulated perturbation of the strain (lattice parameter variation) and composition. What we show in the next section is that these assumptions are only valid over a certain range of thicknesses.

4.4.4.3. Analysis of periodic multi-layers with dynamical theory

Modelling with dynamical theory is the most exact way of simulating the scattering from periodic structures. Fewster (1993, 1998) has considered when the kinematical theory breaks down and this will be briefly summarised here and the effects that occur.

We have already established that the peak position of the layer, in this case the "average" layer, cannot be a reliable measure of the composition or strain value because of the dynamical pulling effect, section 4.4.3. So clearly we have to simulate the profile to obtain a reliable value. Also the whole structure should be simulated and not the average layer composition for the most precise work since the equivalent "average" layer peak does not occur at the same position as the periodic structure "average" layer peak. This error is of the order of 4% for a 50x[AlAs (5nm) / GaAs (5nm)] superlattice on GaAs for example. Let us assume now that we obtain some "average" composition then as stated above we have to take account of the difference in lattice parameters and elastic parameters of the individual components in the superlattice. The composition determined is that for the average lattice parameter and will therefore vary with the ratio of the layer thicknesses. Consider the ternary / binary system then the composition we need to include in the simulation is

$$x = \frac{\overline{x}\Lambda a_{ABC}}{t_{ABC}\overline{a}} \sim \frac{\overline{x}\Lambda}{t_{ABC}} \qquad 4.39$$

Since in the dynamical theory we are concerned with inputting the composition and layer thickness we can derive the average lattice parameter

$$\overline{a} = \frac{\Lambda}{n'+m'} = \frac{\Lambda}{\left\{\dfrac{t_{BC}}{a_{BC}} + \dfrac{t_{ABC}}{a_{ABC}}\right\}} \qquad 4.40$$

where n' and m' are non-integer and the a parameters refer to the direction normal to the layers. The simulation of the rocking-curve on the basis of this "average" layer compared with the simulation of the full superlattice structure is given in figure 4.32. Clearly for the most precise work we have to be aware of the errors introduced in working from some average composition.

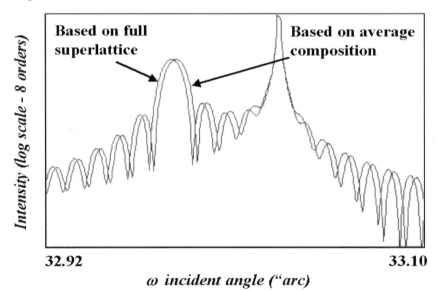

Figure 4.32. The difference in simulated profiles from the average of the superlattice structure and the actual superlattice structure.

Chapter 4 A Practical Guide to the Evaluation of Structural Parameters   237

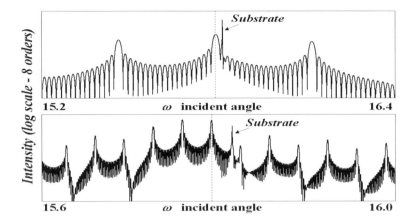

Figure 4.33. As the period increases the simple assigning of the "average" superlattice peak becomes difficult. Both these simulations have the same average composition, yet the longer period structure (lower picture) gives the nearest low angle peak at the position of AlAs instead of the $Al_{0.5}Ga_{0.5}As$ position as in the upper profile.

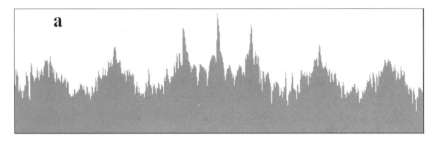

Figure 4.34. The influence of random period variations (a) compared with decreasing periods as a function of depth (b), for the same average period throughout the structures.

Perhaps a more dramatic effect occurs when the period is gradually increased. The concept of an average peak disappears, figure 4.33. Also for very few periods the satellites become broadened and the characteristic length should be determined from the minima and not the maxima since this becomes equivalent to a simple interference problem, section 2.7.2. Other features to note are the complex interference phenomenon in some of the profiles; these can shift the satellite peaks positions at the 1% level. Again this may not be significant for the precision of the analysis required but it is important to be aware of the limit of using a more direct analysis. Clearly modelling the scattering with dynamical theory is the most exact method, however the amount of information required to obtain an exact fit can be more readily found by a feeling of the sensitive features that are more easily observed from a kinematical theory viewpoint.

The variation in the period that was discussed in section 4.4.4.2 can also be confirmed using the dynamical theory simulation, figure 4.34. The case for large variations has been given for a random variation and a systematic variation. The satellites do increase in width with satellite order for both types however the shapes of the satellites are very different, the random variation are more Gaussian, whereas the systematic variation is much squarer in shape. These calculations do not include randomness laterally, which would smear the high frequency fringes.

We can now give some basic guidelines for the applicability of the kinematical and dynamical theories.
- For >3 periods of 4nm<Λ<40nm the kinematical theory (using the integrated intensities of the satellites) gives reliable results.
- For <3 periods or Λ<4nm and Λ>400nm the dynamical theory should be used
- The direct interpretation is a good guide to subsequent simulation or for deriving approximate working values.

These are very approximate guides and in general should be simulated to ensure that these regions are valid for the problem to be solved.

4.4.4.4. Analysis of periodic structures with reflectometry

Reflectometry is sensitive to the composition through variations in the structure factor F_{000} as a function of depth below the surface, section 2.8. The sensitivity is poorer than the experiments close to Bragg peaks since reflectometry is much more dependent on the intensity rather than the angular displacement (strain effect). However reflectometry can prove very useful for the determination of the period in superlattice structures and interface quality providing care is taken in the measurements. Whereas in the previous measurements the diffuse scattering due to imperfections is generally small, close to the *000* origin of reciprocal space (the reflectometry region) these effects can be large and can influence the interpretation. Therefore a second scan is often carried out after the specular scan with a small offset in crystal rotation angle, ω, to determine the diffuse scattering for subtracting from the specular scan. However a bent sample can cause problems with this procedure if the illuminated area varies as with a fixed incident beam divergence, since the broadening in ω will vary as the range of incident angles is reduced with increasing angle. This situation is partially overcome by using variable divergence slits, section 3.7.4.1.

If the quality of the material is very good as in nearly perfect epitaxy then multiple-crystal optics can be used and the profile can be placed on an absolute angular scale. This can aid interpretation. Of course from Chapter 2, we could see that the refractive index is larger for smaller scattering angles and this will influence the satellite positions as well as the fringe spacing effects discussed above. However the direct interpretation, section 4.4.2.2 clearly suggests that simulation of the profile is the only reliable method to interpret the profile.

4.4.4.5. Analysis of a nearly perfect epitaxial periodic multi-layer

In many ways the analysis of a nearly perfect periodic multi-layer does not differ from the multi-layer structures described in section 4.4.3 and the whole analysis can be approached with dynamical theory. For ease of analysis we can assume that each repeat unit is identical and hence the layer thickness can be linked, unless we include some systematic or random variation in the period. We will consider three analyses using the kinematical, optical and dynamical theories.

4.4.4.5.1. Analysis based on the kinematical approach:

The positions of the satellites are defined by equation 4.37 above and the intensity is given by the kinematical equation 2.89 with a correction for the absorption in 2.98. The constant of proportionality in equation 2.89 is effectively the overall scale-factor and this has to be determined. The structure factor can therefore be directly related to the measured intensity. Now the structure factor in this case is as defined in equation 2.22 but in this case the unit cell corresponds to the repeat unit. We can therefore add in aspects of interfacial grading by changing the proportion of partial occupancy, whereas the incommensurability is included by coherently adding several structure factors corresponding to different unit cell (period) combinations as expressed in equation 4.32. The measured intensity should correspond to the integrated intensity after background removal. The process of fitting the data of experimental intensity to the calculated intensity now becomes an iterative process, but unlike the dynamical model we are unconcerned with the profile shape and therefore the number of data points to be matched is related directly to the number of satellites measured.

To simplify the process the number of parameters should be reduced or at least estimated for rapid convergence. The average period is known and should represent the centre of some (e.g. Gaussian) distribution of periods, figure 4.31b. The width of this distribution can be determined from the period variation, equation 4.35. If the lattice parameter of the whole structure is constant parallel to the interfaces throughout its depth then we have just thickness and composition to include throughout one period. If relaxation does exist then the modelling requires a combination of reflectometry and this analysis to achieve a complete picture, Birch, Sundgren and Fewster (1993). In this example we will assume the layers are not relaxed. We should consider the analysis of the interfaces through an understanding of the growing surface, figure 4.35. During deposition it is highly likely that the composition is switched before an exact number of unit cells or atomic layers is complete, also the low surface diffusion will limit the ability for all dips and hillocks to interact. Therefore we must expect some roughness which when averaged laterally appears as a spreading in the interface. The interface will also in general not correspond to an integer number of building blocks (unit cells or atomic layers) therefore we have an incommensurate structure and this should be accounted for by including several combinations (with a minimum of two for each interface).

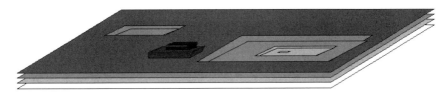

Figure 4.35. The variation in atomic layer completion from limited surface diffusion leading to surface roughening and incommensurate layer structures.

We have now narrowed the problem to a description of the composition (that relates to the strain, through the lattice parameter and Poisson ratio, equation 4.14 to 4.17) within the period. These parameters can now be varied in discrete steps since there are discrete numbers of atomic layers until the best agreement between the experimental and calculated intensity is achieved.

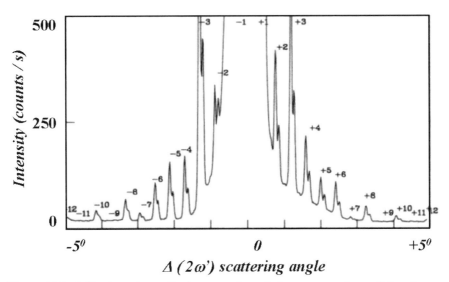

Figure 4.36. A 30 minute data collection scan of a periodic structure using a slit based diffractometer close to the 002 reflection from GaAs. The structure is a GaAs / $Al_xGa_{1-x}As$ superlattice on GaAs.

In this example we shall just consider an AlGaAs / GaAs superlattice and derive the composition and thicknesses of the individual layers and determine the interface extent. Since we will initially be concentrating on the kinematical theory for rapid analysis this method is restricted to the bounds defined in section 4.4.4.3. This example will not use high-resolution instrumentation at all (although there are advantages in doing so) to indicate what can be achieved with what is conventionally called a "powder" diffractometer.

One significant advantage of this approach is that we can use low-resolution diffractometry, section 3.7.1 with all the advantages of high intensity. Also since this is only applicable to periods less than about 40nm the problems of peak overlap are a less significant problem. We can consider the satellites as the coefficients of a Fourier expansion and therefore to obtain good detail about the intermixing at the interfaces we should measure to as higher order as possible. We also should choose a suitable reflection that enhances the satellite intensities and this would depend on whether the contrast between the different layers is dominated by differences in scattering strength or strain. For AlGaAs / GaAs the strain effects are not too strong, however the scattering contrast between them is significant. From equation 4.13 we see that the strain effects become more prominent for higher $2\omega'$ values, whereas the scattering contrast will depend on the relative phases of the scattering components in the structure factor, equation 2.22. For GaAs and AlAs the 002 reflection is dominated by the difference in scattering strength of the Ga and As and the Al and As scattering factors respectively. For GaAs this effect is small, whereas for AlAs this effect is large. This reflection has good contrast and therefore very convenient for (001) surface planes, where a simple scan along $\omega/2\omega'$ (scanning both axes) in the ratio of *1:2* will capture the full diffraction profile.

The data illustrated in figure 4.36 was collected in about 30 min, and after taking account for the background the integrated intensity of each satellite was determined. The full width at half maximum was also measured for each reflection and plotted as a function of satellite order, figure 4.37. The angular positions of all the satellites were also determined. The period was determined by combining all the satellite positions and orders, this also indicated the possible uncertainty in this value, equation 4.28. We now collect data close to the 006 reflection to obtain the strain difference between the "average" superlattice peak and the substrate peak. The high strain sensitivity of this relatively high angle reflection is perfectly adequate for measuring the composition based on the peak separation assumption, section 4.4.1. The lattice parameter difference between GaAs and $Al_xGa_{1-x}As$ for $0<x<1$ is rather small and therefore we can take the simple form of equation 4.29, hence we have two unknowns related by a linear equation (the period and average composition have already been determined above). This is where we now include the intensities of the satellites since these are sensitive to the strength and form of this modulation. In this example a very small interfacial grade was included to determine the individual thicknesses and then the interfacial grade was refined. The results are given in Table 4.3 and the intensity comparison is given in Table 4.4.

Table 4.3. The parameters determined from two quick scans for a simple superlattice structure using the method of direct interpretation and modelling the satellites in the kinematical approximation.

Property	**Value**
Period	25.61(0.1)nm
Period variation	0.53(0.06)nm
GaAs thickness	11.55nm
$Al_xGa_{1-x}As$ thickness	13.46nm
Composition x	0.352
Grade GaAs to $Al_xGa_{1-x}As$	0.54nm
Grade $Al_xGa_{1-x}As$ to GaAs	0.43nm

We could include more complex combinations, three layer repeats and quaternaries, etc., by just cycling over all possible arrangements. In the case of the quaternaries a combination of two or more groups of satellites may be appropriate to find the unique solution.

Table 4.4. The square root of the measured intensities ($F_0 = \sqrt{(I)}$) compared with the calculated structure factor (Fc) from the superlattice, whose parameters are given in Table 4.3. The R-factor for the fit with no grade and for the fit with grading was *18.7%* and *6.6%* respectively, where $R = \dfrac{\sum_N |F_0 - F_c|^2}{\sum_N |F_0|^2}$.

Satellite order	F_0	F_C (no grade)	F_C (grade)
-12	116	304	103
-11	23	7	34
-10	213	361	191
-9	23	121	83
-8	290	410	294
-7	163	293	238
-6	392	443	394
-5	475	566	544
-4	484	455	465
-3	1168	1127	1209
-2	522	415	460
-1	4507	3773	4269
0	Not measured	9779	11140
1	3659	3193	3614
2	571	605	671
3	1016	989	1061
4	567	527	539
5	395	471	452
6	405	468	415
7	147	217	177
8	302	406	292
9	40	62	50
10	193	340	180
11	65	41	37
12	128	272	92

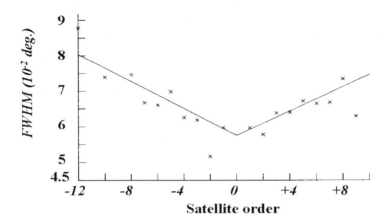

Figure 4.37. The variation of the width of the satellite profiles from figure 4.36 as a function of satellite order. The full-width-at-half-maximum intensity is obtained after stripping the CuKα$_2$ component and fitting to a Gaussian function.

4.4.4.5.2. Analysis based on the optical theory with reflectometry:

The kinematical approach presented above could equally well be used close to the *000* reflection, i.e. from a reflectometry scan. However in this section we will use the optical theory, section 2.8, to model a high quality superlattice with data collected using a multiple crystal diffractometer and an X-ray mirror, figure 3.27 and section 3.8.3. This is an iterative procedure and with the help of direct interpretation or using automatic fitting techniques the fitting can be fairly rapid.

The example in this case is an AlAs / GaAs superlattice to illustrate the sensitivity to the variation in the GaAs to AlAs thickness ratio. The period is very closely determined to the sub 0.1nm level with this many satellites. The measured profile and the best-fit profile are given in figure 4.38, along with the calculated curves for the cases where the GaAs / AlAs thickness ratio has been changed, through thickness increases and decreases of 0.05nm. The differences are quite significant giving a good indication of the sensitivity for this type of superlattice.

The composition sensitivity however is not so good as in the case of scattering close to higher angle reflections. For example changing the composition from AlAs to $Al_{0.9}Ga_{0.1}As$ has a barely perceptible change in the reflectometry profile, whereas close to the 004 reflection the change is very obvious and dramatic. The whole superlattice pattern is shifted with respect to the substrate peak.

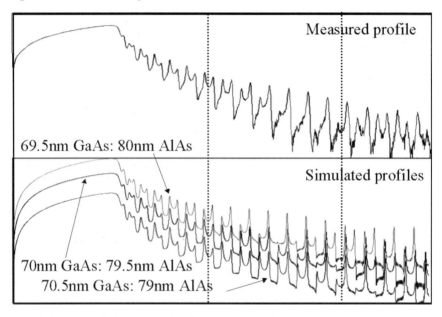

Figure 4.38. The sensitivity of the reflectivity profile to the thicknesses in a periodic multi-layer structures. The best-fit model corresponds to the centre profile of the simulated scans. The data was collected with the multiple-crystal diffractometer.

From section 2.8.2 we can see that any roughening at the interfaces will change the rate of fall in the specular intensity and therefore gives a fairly rapid way of evaluating the interfacial spreading normal to the interfaces. The interfacial smearing normal to the interfaces added in these calculations is 1.27nm (equivalent to 4.5 monolayers) and is consistent with typical values obtained from a large range of AlAs / GaAs samples, Fewster, Andrew and Curling (1991), and the example in table 4.3.

4.4.4.5.3. Analysis based on the dynamical theory simulation:

This approach is very well suited to analysing superlattices with long periods for the reasons discussed in section 4.4.4.3. The procedures are very similar to those given in section 4.4.3.4 for any structure. We shall just illustrate the fit to the structure discussed in the previous section with a single profile and to a reciprocal space map to illustrate the contrasting information. The data were collected with a multiple crystal diffractometer with an X-ray mirror, figure 3.27. The central profile was fitted first using an iterative process and with the full instrumental contribution the measured and calculated reciprocal space maps can be seen in figure 4.39. The agreement is very close with the obvious but very weak diffuse scattering not included. The important aspect to note here even in highly perfect material where an excellent fit from a single profile will show perfect agreement, inspection of the reciprocal space maps illustrates differences. Also it should be noted that the "average" peak for the superlattice does not exist. The most dominant peak in the profile, apart from the substrate peak corresponds to a position for the pure AlAs binary strained to the substrate.

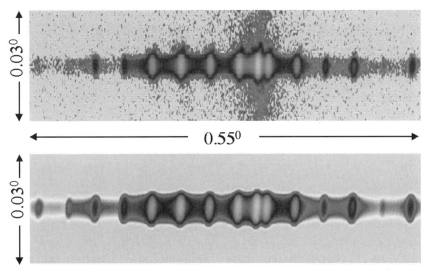

Figure 4.39. The experimental and simulated reciprocal space maps of a near perfect AlAs / GaAs periodic multi-layer structure. The simulation of the diffuse scattering is not included in this model. The scale is on a log(log(intensity)) scale to accommodate the full dynamic range for display purposes.

4.5. Analysis of mosaic structures (textured epitaxy)

The analysis of mosaic structures is very similar to that discussed for mosaic bulk samples. Generally the mosaicity is associated with relaxed layers with imperfect interfaces, but this is not always the case. Consider the reciprocal space map of an unrelaxed mosaic layer on a mosaic substrate, figure 4.40. There are three distinct peaks that we can associate with the substrate reflection and similarly for the layer. The important aspect to remember is that the alignment is based on bringing the scattering vector from one of the blocks into the plane of the diffractometer. If each block is tilted with respect to the aligned block the others cannot all satisfy this criteria.

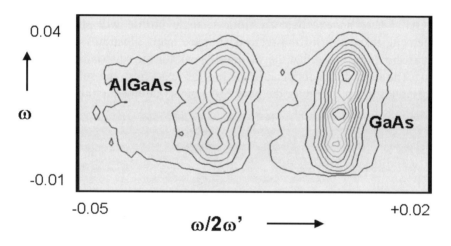

Figure 4.40. A diffraction space map of an AlGaAs layer grown on a mosaic GaAs substrate. The large axial divergence projects all the contributions onto the same plane. A three-dimensional reciprocal space map of the same sample is given in figure 4.12.

If now we undertake a series of maps at different tilt values we can observe this pattern change quite significantly as different blocks move into and out of the optimum scattering condition. If we restrict the axial divergence further we can isolate the scattering from each mosaic block and obtain much improved data and we are able to build up a three-dimensional reciprocal space map of the scattering, figure 4.12. Details of collecting this data are given in section 3.8.3.3. From our reciprocal space map we can see that a simple multiple crystal scan along the surface normal (along

$2\omega'/\omega$ in this case since this is the 004 reflection from a (001) orientated sample) will result in a profile identical to that from a perfect non-mosaic sample. We can therefore apply the simple techniques of interpretation of composition and thickness, etc., given above.

4.6. Analysis of partially relaxed multi-layer structures (textured epitaxy)

As discussed in section 1.6 the growth of thin high quality epitaxial layers is limited by the internal stresses that eventually lead to plastic deformation if they cannot be contained by elastic distortion. From section 2.5, the theoretical description of the scattering clearly also becomes rather complex, see also Kaganer, Koehler, Schmidbauer, Opitz and Jenichen (1997). However we can obtain considerable insight by assuming the scattering behaves in a simple way (i.e. Bragg's equation can be applied directly to any diffraction peak and the correlation lengths can be determined from the real to reciprocal space transforms).

We shall continue our discussion on the influence of dislocations at the interface as discussed in section 4.4.3.8 for nearly perfect structures. If we follow the process during the onset of relaxation then initially the dislocations are isolated with their own strain-field that extends to some limit depending on the stiffness of the material or region that can absorb this strain, for example the surface. Following the discussion of Kidd and Fewster (1994) we can assume that the dislocation strain-fields form isolated regions that will have a lateral extent related to the thickness of the layer (the top surface of the layer is the strain-relieving boundary). Therefore each dislocation will have nearly identical strain-fields in lateral extent with perfect regions of material between them. We therefore have a distribution of correlation lengths associated the shape of the dislocation strain-field and the perfect crystal. For the very early stages of dislocation formation the former is small (comparable to the layer thickness) and the latter is large. We can therefore assume that the scattering is independent so that we can simply add two shape functions, as in figure 2.20, with different length scales.

The result is shown experimentally with the simulated reciprocal space map in figure 4.41. As the thickness of the layer increases the thickness of the defect region enlarges and narrows the diffraction broadening, however the region between defects decreases so that the width of the layer peak (the longer length scale) will increase. Eventually the defect density increases to such an extent that the strain-fields overlap and the extent of the strain-field normal to the interface is restricted and therefore contained close to this interface. The distortion will lead to defects breaking up the structure locally and will lead to tilts and twists. The consequence of this is that the layer peak can exhibit broadening from the limited length scale and due to the different shaped regions having different orientations and lattice plane distortions. For a more rigorous analysis the whole strained region should be sectioned into regions of roughly constant distortion and rotation or considered as a complete scattering unit through equation 2.22. In the limit the defects will completely relieve the internal stresses in the layer and the layer will become totally relaxed, i.e. the layer lattice has the dimensions of that for bulk material of the same phase composition. Of course we should not consider that the distortion is entirely within the layer and assume the substrate is completely unaffected. In the following determination of the relaxation and composition we will take the substrate distortion into effect and then consider how this information can be obtained with different assumptions. We shall then consider how we can begin to extract information on the microstructure.

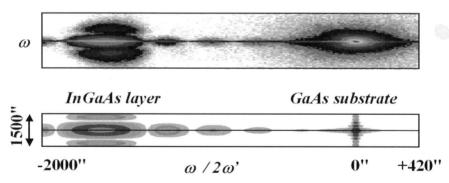

Figure 4.41. The experimental and calculated reciprocal space maps of an $In_{0.1}Ga_{0.9}As$ layer on a GaAs substrate. The layer was just beginning to relax and the diffuse scattering, associated with isolated dislocations close to the layer peak can just be observed. The simulation includes several columns of 70±10nm wide and 70nm thick with a distribution of distorted regions at the interface ranging from 5 to 20nm. The total distorted area was ~10% of the total area and the perfect intervening regions were ~700nm wide.

4.6.1. Measuring the state of strain in partially relaxed thin layers

The state of strain is fundamental to deriving the composition in thin layers. In partially relaxed layers the state of strain cannot be determined from the equations relevant to those for nearly perfect epitaxy, section 4.4. The interatomic spacings above and below a partially relaxed interface are not equivalent and therefore have to be determined independently. We can define the interatomic spacings in the layer and substrate as an orthogonal set d_x and d_y in the plane of the interface and d_z normal to the interface plane. The determination of these spacings will be described with reference to figure 4.42, which is a subset of figure 3.4a when we consider regions close to two layer reflections and two substrate reflections.

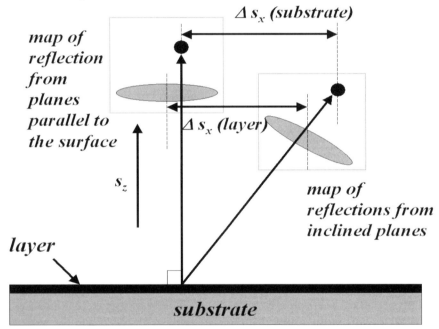

Figure 4.42. A schematic of the measurements to be extracted from two reciprocal space maps to determine the relaxation of a layer on a substrate.

The observation of any reflection can be characterised by an incident and scattering angle, i.e. ω and $2\omega'$ respectively, after the scattering plane normal has been brought parallel to the diffractometer plane with ϕ and χ. The interatomic spacing normal to the sample surface is given by the reciprocal of the differences of two measured positions $\Delta s_z = s_{z1} - s_{z2}$ in equation 3.1

$$d_z = \frac{(n_1 - n_2)}{n_1 \Delta s_z} = \frac{\lambda}{[\sin\omega_1 + \sin(2\omega'_1 - \omega_1)] - [\sin\omega_2 + \sin(2\omega'_2 - \omega_2)]}$$

4.41

For the interatomic spacings in the plane of the interface we can write

$$d_x = \frac{(n_1 - n_2)}{n_1 \Delta s_x} = \frac{\lambda}{[\cos\omega_1 - \cos(2\omega'_1 - \omega_1)] - [\cos\omega_2 - \cos(2\omega'_2 - \omega_2)]}$$

4.42

and similarly for d_y.

Hence if we measure the angular positions of two reflections for the layer and similarly for the under-layer, or substrate, then we can determine the difference in the interplanar spacing parallel and perpendicular to the interface.

For two reciprocal lattice points separated by Δs_z It is clear that if we know the angles ω and $2\omega'$ on an absolute scale then we are defining the location of the origin of reciprocal space with Miller indices 000 and we should set $\omega_2 = 0$ and $2\omega_2' = 0$. We can therefore determine the interplanar spacing parallel and perpendicular to the surface using one reflection for each azimuth. The determination of the $2\omega'$ angle on an absolute scale is relatively easy, section 4.3.4. The ω absolute angle is less precise, in general, however since this is a relative angle with respect to the surface it will bring in $\cos(\omega)$ errors to the measured interplanar lengths and for small errors this will not be significant. This angle basically just defines what direction is perpendicular to the surface and only becomes significant when this is related to the assumed orientation associated with the phase composition. We can simply overcome this by defining the perpendicular and parallel interplanar spacings along an appropriate direction, e.g. relate it to the [001] direction for a nominal [001] orientated sample.

4.6.2. Obtaining the composition in partially relaxed thin layers

This section relies on the determination of the interplanar spacings parallel and perpendicular to the interface of interest; the general principles are given above and the various practical methods are given in sections below. However now we should refer to equation 4.17, but modify it slightly so that the interplanar spacings are with respect to the layer in the fully relaxed state

$$x = \frac{\{d_M\}_{Relax} - d_B}{d_A - d_B} \qquad 4.43$$

where d are suitably chosen parameters characteristic of the composition, i.e. the unstrained (fully relaxed) interplanar spacings for the alloys forming the layer. In the cubic case, as discussed previously, this could be the unit cell dimension, a, whereas in the hexagonal case this could be c, and similarly for any other symmetry. The subscripts A and B represent the phase limits of the alloy A_xB_{1-x} and $\{d_M\}_{Relax}$ is the determined interplanar spacing along the same direction when all the strain has been relieved, i.e. the equivalent bulk value. Equation 4.43 assumes Vegard's rule however whatever the relationship between composition and interplanar spacing the same principle applies. For example equation 4.18 accounts for a small modification to this rule, yet there is a simple relationship between composition x and a lattice parameter for the $Si_{1-x}Ge_x$ alloy system.

Our concentration now is to determine $\{d_M\}_{Relax}$. From the previous section we will assume that we have determined the interplanar spacing in three orthogonal directions (normal and in the plane of the interface). Suppose now that the interplanar spacings in these three directions for the two limits of the alloy extent are related by certain constants, such that

$$\begin{aligned} _1d_x :_1 d_y :_1 d_z &= {}_1K_x :_1 K_y :_1 K_z \\ _2d_x :_2 d_y :_2 d_z &= {}_2K_x :_2 K_y :_2 K_z \end{aligned} \qquad 4.44$$

These K parameters relate to the choice of the reflection in the measurement, e.g. a choice of reflections from a cubic [001] orientated layer may be 004 (along z), 444 (along x) and -335 (along y): this will give

$K_z = 4$, $K_x = \sqrt{(4^2+4^2)}$ and $K_y = \sqrt{(3^2+3^2)}$. We will have to assume that these K parameters vary for the alloy in a well-defined manner and it is not unreasonable to assume that they follow the same relationship with composition as the interplanar spacings (i.e. Vegard's rule in general). For the cubic case all these K values are unchanged and for hexagonal structures these will follow the c/a ratio that can vary between phases.

The d values in equation 4.44 are all the fully relaxed values and therefore we must derive these from the strained values and the appropriate elastic parameters. We shall concentrate on the interplanar spacing normal to the interface and derive the $\{d_z\}_{Relax}$, now

$$\varepsilon_{xx} = \frac{d_x - \{d_x\}_{Relax}}{\{d_x\}_{Relax}}$$

and similarly for ε_{yy} and ε_{zz}. Combining these relations, equation 4.44 and equation 1.7, we obtain after some manipulation

$$\{d_z\}_{Relax} = K_z \frac{1-v}{1+v}\left\{\frac{d_z}{K_z} + \frac{v}{1-v}\left(\frac{d_x}{K_x} + \frac{d_y}{K_y}\right)\right\} \quad 4.45$$

v is the Poisson ratio for the phase along the direction normal to the surface. This is unknown (since the alloy composition is unknown at this stage) although we know the values at the limits of the two phases. So we relate the Poisson ratios to the composition

$$v_{A_xB_{1-x}} = xv_A + (1-x)v_B \quad 4.46$$

If we initially assume the Poisson ratio as the mean value, i.e. for $x = 0.5$, then a first estimate of $\{d_z\}_{Relax}$ can be used to determine a composition with equation 4.43. We can then refine the Poisson ratio iteratively by substituting the new value of x into equation 4.45 until the Poisson ratio change is insignificant. Similarly the K values will follow the same relationship and should be included in the iterative process.

4.6.3. The measurement of the degree of relaxation and mismatch in thin layers

These parameters can be useful for estimating the number of stress relieving defects. Until now we have treated the substrate and layer independently because the state of strain gives the composition, etc., whereas the mismatch and degree of relaxation just relate the strain in the layer to that of the substrate or layer below. The layer strain along x in the plane of the interface is

$$Strain = \varepsilon_{xx} = \left\{ \frac{d_x - \{d_x\}_{Relax}}{\{d_x\}_{Relax}} \right\} \qquad 4.47$$

and similarly for ε_{yy}. This strain should be considered in two orthogonal directions to fully characterise the misfit at the interface. The strain normal to the interface is given by

$$Strain = \varepsilon_{zz} = \left\{ \frac{d_z - \{d_z\}_{Relax}}{\{d_z\}_{Relax}} \right\} \qquad 4.48$$

The degree of relaxation is simply the ratio of the difference in the actual interplanar spacing parallel to the interface to that for the layer in the completely relaxed state.

$$R_x = \frac{d_x - {_s}d_x}{\{d_x\}_{Relax} - {_s}d_x} \qquad 4.49$$

${_s}d_x$ is the interplanar spacing along the direction x (parallel to the interface) for the layer below that of interest or the substrate. We can similarly write an expression for R_y. The value of the relaxation in general will vary from 0 (no relaxation: the layer is perfectly strained to fit the under-layer) to 1 (fully relaxed: the strain has been completely relieved). This relaxation value is a convenient simple input for modelling the strain to be included in the simulation of the scattering profile. The strain is determined from the relaxation parameter through equations 1.8 and 1.9.

The defect density can be derived from the difference in the interplanar spacing parallel to the interface. If all the defects have the same contribution to the strain relief and the component of their Burgers vector parallel to the interface is known then from equation 4.26 we can determine their density. Of course we must be aware that the defects can pile up especially at high levels of relaxation or a mixture of dislocations types may exist bringing doubt into the estimation.

We shall cover various approaches to the determination of composition and relaxation including the use of reciprocal space maps, rocking curves, etc., and compare their precision with examples.

4.6.4. *The determination of relaxation and composition with various methods*

In this section we shall give examples of structures that are partially relaxed that also include some tilting and analyse them using various options. The most reliable method will be to interpret the data using reciprocal space maps and calculating the parameters from the absolute angles determined from the peak positions. We can then make some assumptions and gradually work our way through the procedures to single peak determination of composition of relaxed structures. Collecting reciprocal space maps should not necessarily be considered a slow process and results obtained from data collection in a few minutes are perfectly useable. Rocking-curve measurements, assuming that the substrate lattice parameter is known can be used as an internal standard and will also be considered.

4.6.4.1. Determination by reciprocal space maps on an absolute scale:

This method is suitable for a multiple crystal diffractometer with a monochromator and an analyser. Suppose we wish to obtain the most precise value for the parallel and perpendicular interplanar spacings then we should measure combinations of reflections that have the largest separations in s_x, s_y and s_z. This is purely a geometrical consideration. However we must include some indication of the volume of the sample that is being assigned to these determined parameters. The X-rays penetrate the sample and are attenuated depending on the general photoelectric absorption and extinction effects; these topics are discussed in detail in Chapter 2.

Generally if we are determining the parameters in a thin layer, then we sample the whole layer and obtain a reliable average. The substrate will almost certainly be distorted close to the interface due to the strain-field of the dislocation, yet deep below the interface the parameters will be representative of the bulk parameters. To obtain the interplanar spacings of the substrate, in order to extract the degree of relaxation, we should compare results from reflections with a similar probing depth, which should ideally be optimised to be close to the interface. For example for a [001] orientated GaAs wafer nearly all the intensity is scattered in the top 5μm for the 004 reflection and similarly for the 444 reflection, however for the 224 reflection this depth is 2.7μm, Fewster and Andrew (1998). These depths are determined by simulation and noting the rate of change in the integrated intensity as a function of thickness (effectively regions of the substrate below depths that have little effect on the integrated intensity are not contributing significantly to the scattering).

Let us now consider the accuracy of determining the interplanar spacing for different reflections. As discussed in section 4.6.1, the $2\omega'$ angle can be determined with high precision (i.e. the zero can be set precisely) whereas the ω angle is at best an order of magnitude less precise since this relates to the sample surface. Therefore if we wish to determine the strain perpendicular to the interface the rotation in ω should be normal to this direction for the reflection we use; hence scattering from planes parallel to the surface satisfy this condition, therefore the 002, 004, 006, etc., for a [001] orientated surface are best. We should now apply our criteria of largest separation in s_z, since we are on an absolute scale we use the chosen reflection in combination with the 000 "reflection." Remember of course we can arbitrarily define the direction of s_z along the main plane normal of the substrate, without introducing too much error and this can define the ω angle zero. We therefore can write equation 4.41 on an absolute scale as

$$d_z = \frac{\lambda}{[\sin\omega + \sin(2\omega' - \omega)]} \qquad 4.50$$

To determine the interplanar spacing parallel to the interface the errors introduced by the uncertainty in ω are minimised for reflections from planes normal to the surface. We have described how these in-plane measurements are conducted; section 3.8.3.5 and these are generally unsuitable for any rapid method. Complications do arise also because this will give a projected angle. However since we are concerned about the uncertainty in ω, from our measurement of d_z we have now defined ω and with a precision goniometer this is now as precise as the $2\omega'$ determination, although the absolute $\omega = 0$ position is uncertain. Therefore a combination of any two reflections with differences in s_x will give the interplanar spacing parallel to the interface (or more precisely along a direction normal to s_z). If we measure the 444 reflection with respect to the 004 reflection for a [001] GaAs sample then to maintain 1ppm accuracy in the interplanar spacing parallel to the interface (assuming a perfect measurement) ω should be accurate to within 0.1^0 of the absolute value. We therefore obtain d_x from equation 4.42, where the subscript *1* refers to the 444 reflection and *2* to the 004 reflection for example. To account for any anisotropy the measurement should be performed along both the x and y directions. The d values are then substituted into equation 4.45 and with equation 4.43 the composition is obtained.

The relaxation along the x and y directions are determined with equation 4.49, since all the components are known. If we can make assumptions concerning the relaxation per defect then the interfacial defect density can be derived from equation 4.26 with the appropriate Burgers vector contribution.

4.6.4.2. Determination by using a series of rocking curves and analyser scans

This method is suitable for a multiple crystal diffractometer with a monochromator and an analyser. It is clear from the above analysis that we only need the ω and the $2\omega'$ values for two reflections and the clearest method is to undertake a reciprocal space map including the layer and the substrate, or some other layer reference reflections. Now of course we can obtain the ω value directly from the rocking curve, provided that the layer peak is well defined and the ω reference is defined as described in section 4.6.4.1. If we drive the ω rotation to the layer peak position then with this open detector the instrument acceptance is given as in figure 3.30a. If we invoke the analyser, which is already defined on an absolute scale then a simple $2\omega'$ scan will give the second angle we need. The procedure then simply follows that given in the previous section.

This method is less accurate than conducting reciprocal space maps because an inaccurate location of the ω peak position, which can be very broad, will introduce additional errors in the $2\omega'$ value. Consider the example of a heavily relaxed GeSi layer on Si, where the rocking curve gives a very poorly defined layer peak, figure 4.43a. In contrast the reciprocal space map gives a clear peak for a good determination of the ω angle, figure 4.43b. A perfectly measurable reciprocal space map can be obtained in 5 minutes with the multiple crystal diffractometer, 220 symmetric reflections throughout and line focus Cu X-ray source with mirror. Clearly the reciprocal space map is the most reliable approach when the layer peak is broad and flat. However for reasonably well defined layer peaks this can be a very quick and useful approach.

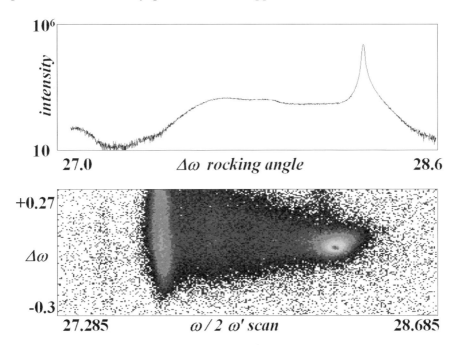

Figure 4.43. The difficulties associated with determining the ω angle from a rocking curve compared with a reciprocal space map for a heavily relaxed SiGe layer on a Si substrate.

4.6.4.3. Determination by reciprocal space maps on a relative scale:

This method relies on the substrate, or some other internal reference, to have known ω and $2\omega'$ angles. The layer reflection of interest is then collected on the same map and the ω and $2\omega'$ angles are determined very simply from

$$\omega = \omega_S + \Delta\omega$$
$$2\omega' = 2\omega'_S + \Delta2\omega$$

4.51

The procedure then follows that given above, as though this information was determined on an absolute scale. It must of course be remembered that the reference reflection may not be representative of the substrate material deep below the interface. The substrate or reference under-layer may be strained due to the interface distortion and consequently the resulting composition and degree of relaxation may well be in error. This is best shown with an example, Fewster and Andrew (1998).

Table 4.5. Composition and relaxations in a thick InGaAs layer on GaAs; determined assuming the lattice parameter of the substrate is a reliable internal standard and then determining the same values on an absolute scale.

Method	a_z	R_x (%)	a_x	R_y (%)	a_y	$a_{unstrained}$	x (%)
Relative							
Substrate	5.65368		5.65368		5.65368		
Layer	5.67546	108	5.67420	63	5.66550	5.6728	4.75
Absolute							
Substrate	5.65367		5.65386		5.65397	5.65386	
Layer	5.67573	90.3	5.66978	90.7	5.66996	5.67148	4.39

A series of reciprocal space maps from a thick buffer layer of $In_{0.05}Ga_{0.95}As$ grown on GaAs has been analysed assuming the substrate reflection is a true representation of the bulk GaAs lattice parameters. The data was extracted from two 004 reflections for the [110] and [1-10] azimuths and the 444 and 4-44 reflections. The consequent analysis gives a value of composition and relaxation in Table 4.5. The same procedure gives quite a different result if we make no assumptions about the substrate and determine the lattice parameters on an absolute scale. Clearly the degree of relaxation changes from a value that is above 100% in one direction to something more realistic as we go from the relative method to the absolute method. The composition is also different when determined by the two methods.

4.6.4.4. Determination by rocking curves alone

This method is suitable for instruments with a monochromator and open detector or a double crystal diffractometer. A rocking curve just gives the difference angle $\Delta\omega$ and therefore we have to find some way of obtaining the 2ω' angle for each reflection. This method relies on the substrate reflections having known ω and 2ω' angles.

Suppose initially we consider the layer and substrate to be perfectly aligned with each other, i.e. there is no tilt, then

$$\left\{\frac{\Delta s_z}{\{s_z\}_S}\right\}_H = \left\{\frac{\Delta s_z}{\{s_z\}_S}\right\}_0 = const \qquad 4.52$$

where the subscript H refers to an arbitrary reflection and 0 to a reflection from planes parallel to the surface. $\Delta s_z = s_z - \{s_z\}_S$ where s_z refers to the layer and $\{s_z\}_S$ refers to the substrate. This assumption clearly relates to the fact that $\Delta d/d$ is constant along a given direction. From equation 4.41 we can state that

$$\{\Delta s_z\}_H = \left\{\frac{\Delta s_z}{\{s_z\}_S}\right\}_0 \{s_z\}_H = \sin\omega_L + \sin(2\omega'_L - \omega_L) - \sin\omega_S - \sin(2\omega'_S - \omega_S)$$

$$4.53$$

Now since we are dealing purely with relative positions there is doubt about the surface normal direction and this has to be defined or chosen to be parallel to a convenient substrate scattering plane normal. We can determine the tilt of the layer with respect to the substrate by collecting 2 rocking curves at azimuths 180^0 apart in ϕ.

$$Tilt = \varphi = \frac{1}{2}\{\Delta\omega_0 - \Delta\omega_{180}\} \qquad 4.54$$

Equation 4.53 then should be

$$\{\Delta s_z\}_H = \left\{\frac{\Delta s_z}{\{s_z\}_S}\right\}_0 \{s_z\}_H = \sin\{\omega_L + \varphi\} + \sin(2\omega'_L - \{\omega_L + \varphi\}) - \sin\omega_S - \sin(2\omega'_S - \omega_S)$$

$$4.55$$

Rearranging this equation we obtain the $2\omega'$ angle for the reflection of interest; the subscript, L, to this layer reflection has been omitted.

$$2\omega' = \sin^{-1}\left\{\left(\frac{\Delta s_z}{\{s_z\}_S}\right)_0 s_z - \sin\{\omega + \varphi\} + \sin\omega_S + \sin(2\omega'_S - \omega_S)\right\} + \omega + \varphi$$

$$4.56$$

Therefore we have all the parameters necessary to determine the scattering angle $2\omega'$ and ω is found through equation 4.51 and because we are defining the surface normal perpendicular to a set of substrate scattering planes, i.e. $2\omega'_S = 2\omega_S$ from equation 4.55:

$$\left(\frac{\Delta s_z}{\{s_z\}_S}\right)_0 s_z = \sin\{\omega + \varphi\} + \sin\{\omega - \varphi\} - 2\sin\omega_S \qquad 4.57$$

A combination of a rocking curve from a set of planes inclined to the surface plane for $\varphi \neq 0^0$ and two rocking curves for $\phi = 0^0$ and 180^0 from planes parallel to the surface will then give us all the information to obtain the composition and relaxation, etc., along the x azimuth. To account for any anisotropy we should include a similar set along y. Additional precision can be achieved by including widely spaced reflections from inclined planes as well as from planes parallel to the surface. All the discussion concerning problems with relative measurements must be born in mind. It should also be clear that this method cannot achieve the accuracy of the reciprocal space map method, as demonstrated in Table 4.5.

4.6.4.5. Revealing dislocations and defects by topography:

The above methods of determining the relaxation in thin layers are only suitable for large relaxation. When the density of dislocations is small, as in the sample described in section 4.4.3.6, then we can count the number and obtain the dislocation density. However it is sometimes useful to take topographs of samples when the defect density is clearly well above the level to observe single dislocations.

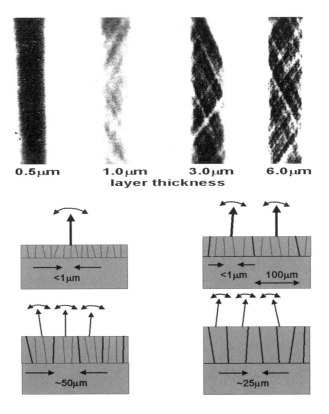

Figure 4.44. (a) The evolution of the dislocation network in $In_{0.05}Ga_{0.95}As$ on GaAs as a function of layer thickness and (b) the proposed model of the microstructure based on these topographs and the relaxation measurements.

The microstructure of materials at various stages of relaxation can reveal the nature of process. We shall consider a series of $In_xGa_{1-x}As$ layers, where $x \approx 5\%$. As discussed previously the present level of resolution achievable with topographs is related to the developed emulsion size (approximately 1 micron, although the diffractometer resolution may increase this value). In figure 4.44a there are four topographs taken from four different thickness of layers with the same intended *In* composition. It is clear that the contrast increases with layer thickness. This can be interpreted as the amalgamation of very small regions with large tilts to create larger regions with similar tilts suggesting a mosaic grain growth, rather similar to work hardening, Fewster and Andrew (1993c). The characteristic length scales and interpretation is given in figure 4.44b. This example combines the analysis of reciprocal space maps with topography and indicates the tools available for this sort of analysis in understanding the evolution of the microstructure.

4.6.4.6. Simulating structures with defects

In section 2.5 we described an extension to the dynamical model to take into account the influence of defects at interfaces. The strain associated with the misfit dislocations and associated defects will extend into the layer above and below and at low levels this can be characterised by the shape of the diffuse scattering close to the substrate and layer peaks. At high levels this can be very pronounced and can lead to significant diffuse scattering streaks parallel to the interface plane. This example indicates the process of extracting information by modelling the full profile.

The sample was a 3μm $In_{0.05}Ga_{0.95}As$ layer on a GaAs substrate. On top of this layer there was a superlattice of alternating $In_{0.10}Ga_{0.90}As$ and GaAs layers, whose average lattice parameter should be matched to the layer below. The materials problem was to ascertain the strain in the superlattice, was it isotropic, etc. This required a full analysis to ensure that the relaxation in the 3μm layer was complete and the material quality was reasonable. The method of determining the relaxation was carried out as described above, section 4.6.4.3 and this was included in the simulation. The data were collected with a limited-area reciprocal space map close to the 004 reflection and projected onto the *[001]* direction. This recovered the full intensity associated with tilting and limited mosaic size effects and captured the diffuse scattering associated with the interfaces.

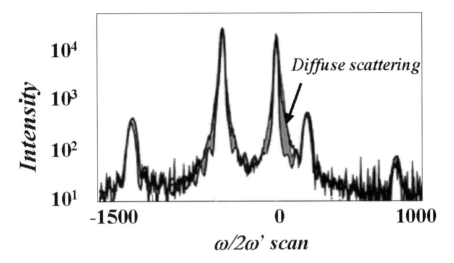

Figure 4.45. The profile from projecting a diffraction space map of a 3μm layer of $In_{0.05}Ga_{0.95}As$ on GaAs with a $In_{0.1}Ga_{0.9}As$ / GaAs superlattice on top. The inner profile is the best fit based on a structure being perfect, although the best fit to the experimental profiles, with all the diffuse scattering included, gave the extent of the interfacial distortion.

Since the degree of relaxation was well above the few percent level the X-ray wave-fields are uncoupled at the interface and this was shown to immediately influence the relative heights of the layer and substrate peaks. Matching peaks alone assuming the material is perfect gave a thickness value of 2μm, however when the wave-fields are uncoupled in the simulation the value was close to 3μm. The diffuse scattering contribution alters the shape of the peaks at the base of the substrate and layer peak. By including the strain distribution this can be modelled. This region is divided into regions of constant strain which are considered to scatter independently (the exponential variation in the strain is split into 10 layers above and below the interface). The shape is then fitted in an iterative manner until the best fit is found. The additional variables to model the diffuse scattering are purely the extent of the strain-fields above and below the interface, figure 4.45.

4.7. Analysis of laterally inhomogeneous multi-layers (textured polycrystalline)

In the example above we have taken a reciprocal space map from planes parallel to the interface and projected it onto a direction normal to the interface. Although we can achieve a good fit to the profile there is a considerable amount of information that is lost, for example the size of the mosaic blocks, the distance between defects and the distribution of tilts. This will be covered in this section. We shall firstly consider the analysis of lateral inhomogeneities that are too small to be observed by topography and then consider the larger aspects of inhomogeneity. In the first part we will consider a very direct analysis and show how simulation of reciprocal space maps gives a more complete picture before considering topographic methods.

4.7.1. Direct analysis of laterally inhomogeneous multi-layers:

From section 2.7.3 we derived some expressions for the influence of finite size regions on the scattering pattern. Also any region that is tilted with respect to the average will add to the broadening and consequently we have a combination of contributions. If we assume that the "mosaic blocks" are perfect and their bounds are primarily limited parallel to the surface plane then we can consider the various components that contribute to the shape and position of the measured scattering, figure 4.46a.

Scattering from planes approximately parallel to the surface will have overlapping contributions from lateral correlation lengths (lateral finite sizes) and the distribution of tilts (microscopic tilts) giving a characteristic broad elliptical scattering shape. For scattering from planes inclined to the surface plane these contributions will be inclined to each other and will rotate the diameter of the ellipse depending on the strength of the various contributions. Consider figure 4.46b, a schematic of a reciprocal space map, where the contribution from the lateral correlation length, L_1, is parallel to the surface plane and the microscopic tilt contribution, L_2, is normal to the radial direction (along an $\omega/2\omega'$ scan direction). The full-width-at-half-maximum intensity of the overall ellipse can be characterised by two lengths in reciprocal space, i.e. Δs_x and Δs_z. Now the two contributions will be correlated and we can consider them as vector sums of the microscopic tilt perpendicular to the radial direction, L_2, and the lateral correlation length parallel to the surface plane, L_1.

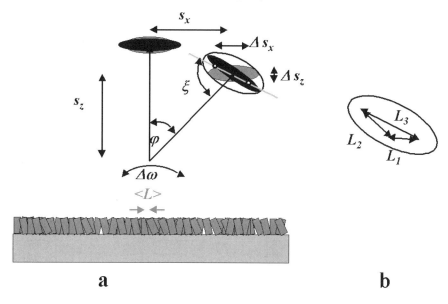

Figure 4.46. (a) The structural features that give rise to the shape of the scattering in reciprocal space are the microscopic tilt distribution black ellipse and lateral correlation length grey ellipse. (b) gives the proportion of these components with respect to the whole general shape.

Hence from trigonometry we obtain

$$\frac{L_2}{L_1} = -\frac{\cos\xi}{\cos(\varphi+\xi)} \qquad 4.58$$

Similarly we can write

$$\frac{L_3}{L_2} = \frac{\sin\varphi}{\cos\xi} \qquad 4.59$$

In reciprocal space units L_3 is simply given by

$$L_3 = \sqrt{(\{\Delta s_x\}^2 + \{\Delta s_z\}^2)} \qquad 4.60$$

and

$$\varphi = \tan^{-1}\left\{\frac{s_x}{s_z}\right\} \qquad 4.61$$

$$\xi = \tan^{-1}\left\{\frac{\Delta s_x}{\Delta s_z}\right\}$$

Hence

$$Lateral\ correlation\ length = \frac{1}{L_1} \qquad 4.62$$

and the microscopic tilt (in radians) is related to the dimension L_2 by

$$Microscopic\ tilt = \frac{L_2}{\sqrt{\{s_x^2 + s_z^2\}}} \qquad 4.63$$

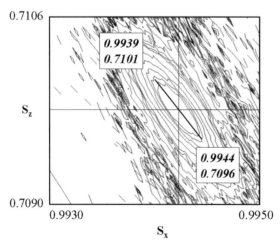

Figure 4.47. The reciprocal space map of a relaxed InGaAs layer (444 reflection). From the alignment of the ellipse and the values of the half maximum intensity positions an estimate of the microscopic tilt (0.03°) and the lateral correlation length (700nm) could be obtained.

As an example consider the reciprocal space map given in figure 4.47. The ellipse is clearly inclined with respect to both the surface normal and radial direction and therefore includes both microscopic tilt and lateral correlation lengths. By measuring the positions of the half-height intensity along the ellipse we can determine the microscopic tilt as 0.03^0 and the lateral correlation length to be *700nm*.

We can extend this method to two reflections from planes parallel to the surface. This could be applicable to materials that are textured with poor orientation preference in the surface plane. The full width at half maximum intensity (FWHM) of two reflections along the same direction in reciprocal space will be denoted by the coordinates as $(\Delta s_x)_1$ and $(\Delta s_x)_2$. The microscopic tilt, $\Delta\omega$, and the lateral correlation length, L, will influence both reflections, n; if we consider both these parameters and denote their contributions by $(\Delta_\omega s_x)_n$ and $(\Delta_L s_x)_n$ respectively then

$$\Delta\omega = 2\tan^{-1}\left\{\frac{\Delta_\omega s_x}{2s_z}\right\}_1 = 2\tan^{-1}\left\{\frac{\Delta_\omega s_x}{2s_z}\right\}_2 \qquad 4.64$$

and

$$L = \left\{\frac{1}{\Delta_L s_x}\right\}_1 = \left\{\frac{1}{\Delta_L s_x}\right\}_2 \qquad 4.65$$

Now as before we can consider the resultant broadening to be the correlation of the two contributions (and the profile are approximated to Gaussian), i.e.

$$\{\Delta s_x\}_n^2 = \{\Delta_\omega s_x\}_n^2 + \{\Delta_L s_x\}_n^2 \qquad 4.66$$

Then from equation 4.64, we have after substituting in equation 4.66

$$\{\Delta_\omega s_x\}_1 = \{\Delta_\omega s_x\}_2 \frac{\{s_z\}_1}{\{s_z\}_2} = \left[\{\Delta s_x\}_2^2 - \{\Delta_L s_x\}_2^2\right]^{1/2} \frac{\{s_z\}_1}{\{s_z\}_2} \qquad 4.67$$

If we now include the relationship given in equation 4.65, along with equation 4.66, we have

$$\{\Delta_\omega s_x\}_I^2 = \frac{\left[\{\Delta s_x\}_2^2 - \{\Delta s_x\}_I^2\right]}{\left[\{s_z\}_2^2 + \{s_z\}_I^2\right]} \{s_z\}_I^2 \qquad 4.68$$

Since we know s_z (compare with figure 4.46) we can derive the microscopic tilt $\Delta\omega$ from equation 4.64 and the lateral correlation length from rearranging equation 4.66 and substituting into equation 4.65

$$L = \left\{\frac{1}{\Delta_L s_x}\right\}_I = \left\{\frac{1}{\{\Delta s_x\}_I^2 - \{\Delta_\omega s_x\}_I^2}\right\}^{1/2} \qquad 4.69$$

This can be a useful approach to the evaluation of these parameters and avoids the difficulty of determining the angle of the ellipse in the former method.

4.7.2. Simulation of laterally inhomogeneous multi-layers:

The analysis above makes several assumptions. The characteristic length scale is assumed to uniform and really takes no real account of the detail of the shape of the profile. The advantages of simulation of the reciprocal space map will be become evident in the following example.

GaN is known to be full of defects when grown on sapphire and the spacing between these defects is an important parameter. The defect separation is analogous to the lateral correlation length. In the first instance we can analyse the reciprocal space map as above and extract some starting values for the microscopic tilt and lateral correlation length. In this example the lateral correlation length with a barely detectable microscopic tilt dominated the shape when analysing the 105 reflection. The derived lateral correlation length is *92nm* and this value was included in the simulation of the reciprocal space of the 002 reflection from the GaN. The GaN is in the hexagonal form with the c-axis normal to the surface and parallel to the c-axis of the sapphire substrate. The multiple-crystal scan along the radial direction ($\omega/2\omega$' scan) is shown in figure 4.48 along with the best-fit profile. The best-fit profile was found only by assuming that the In in the InGaN layer had segregated into the AlN cap region. The agreement was not perfect; in fact the quality of this particular sample was very poor, as we shall see in the next section 4.8.

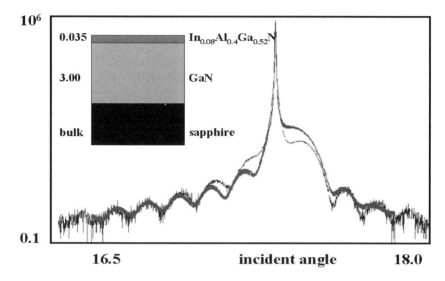

Figure 4.48. The best fit profile to a GaN structure on sapphire. The interface was heavily distorted and the InGaN had diffused into the AlGaN cap layer. The sapphire scattering peak occurs at a much higher incident angle. This 002 reflection was measured with the multiple-crystal diffractometer.

From section 2.6 we can simulate the reciprocal space map based on the determined lateral correlation length and the layer thicknesses and composition determined from the radial scan. The measured reciprocal space map is given in figure 4.49a and the simulated profile in 4.49b. The general shape of the simulated profile is clearly not very close to the measured profile, although the full-width-at-half-maximum intensity for the "substrate" and "layer" peaks are the same. The difference arises because there is a distribution of sizes and this changes the distribution in the tails of profile parallel to the surface. When a distribution is included the appearance of the map changes quite dramatically. Another major contribution to the shape of the diffraction tails arises from the distribution of tilts of the mosaic blocks, equation 2.86 and figure 2.18. So to obtain a good fit to the whole profile some average dimension should be fitted first, followed by a distribution in their sizes and finally the spread of tilts associated with the block wall edges. The best fit is given in figure 4.49c.

272   X-RAY SCATTERING FROM SEMICONDUCTORS

It should be noted that the distribution does not encompass the apparent average derived from the direct analysis in this case, although the full-width-at-half-maximum intensity of both maps is the same. This could have important consequences in deriving quantitative information from these experiments.

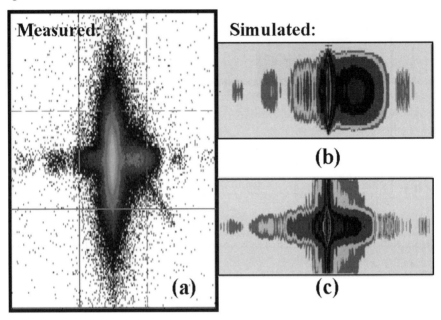

Figure 4.49. The 002 reflection from the GaN structure of figure 4.48; (a) measured reciprocal space map, (b) the simulated reciprocal space map assuming the separation between defects are all similar (92nm) and (c) assuming a distribution of sizes (150±50nm).

4.7.3. Lateral inhomogeneities without large misfits:

In last few examples we have seen that large misfits have led to the breaking-up of the layers into mosaic blocks, however this situation can still exist without relaxation taking place. Examples include mosaic substrates; this is described in section 4.3.2.3 and illustrated in figures 4.11 and 4.12. The layer is seen to mimic the substrate indicating perfect registry between the substrate and the layer within each mosaic block. However each block can be misorientated with respect to each other. For detailed analysis of these structural types full three-dimensional mapping is ideal, section

Chapter 4 A Practical Guide to the Evaluation of Structural Parameters 273

3.8.3.3, or great care must be exercised in the interpretation of the reciprocal space maps since these represent projections of the scattering onto an arbitrary plane.

When lateral inhomogeneities are greater than about 5μm then topography can prove to be a very powerful method of analysis. Again this is best performed in combination with the reciprocal space maps, since this assigns the scattering feature to the "real" space image. Examples of these have been given in previous sections but for completeness is mentioned here. Under some growth conditions twinning can exist and knowing the lateral sizes of the various components can be of great interest. Two examples will be given one for the case when the twins are too small to be observed by topography and the other when they can be observed.

4.7.3.1. Analysing epitaxial layers with very small twinned regions

This is an example taken from Fewster and Andrew (1993b), who analysed a high critical temperature superconducting oxide on a strontium titanate substrate. The layer has orthorhombic symmetry with the *a* and *b* axes of very similar lengths (*0.383nm* and *0.388nm* respectively), which are very close to the SrTiO$_3$ lattice parameter of *0.390nm*. This combination creates the possibility of the *a* and *b* parameters aligning along two directions at 90^0 to each other. This can be envisaged as different nucleation regions during deposition having different orientations and effectively appearing as a distribution of twin components. Using diffractometry the relative proportions can be obtained from the area under these peaks.

Clearly any analysis from planes inclined to the surface normal will have a well defined ω and $2\omega'$ that relates to the lattice parameters and orientation. Therefore scattering within the *0kl* and *h0l* planes for the two possible orientations will have slightly different angular settings when $h = k$ due to the small differences in lattice parameters along *a* and *b*. The chosen reflections for this example were *038* and *308*. The angular settings for $CuK\alpha_1$ radiation for these reflections are $\omega = 4.360^0$, $2\omega' = 105.5506^0$ and $\omega = 4.5873^0$, $2\omega' = 106.891^0$ respectively. The angular separation is reasonable and material quality is poor resulting in weak scattering and therefore high intensity low angular resolution data is the most appropriate.

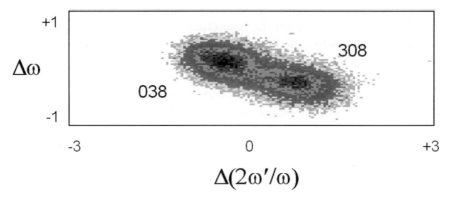

Figure 4.50. The scattering from the 038 planes of one orientation of YBaCuO with the 308 scattering planes of the 90^0 twin component.

The instrument used was a slit-based diffractometer with the configuration described in section 3.7.2. Since no single scan was able to scan through both peaks and create a direct comparison of their intensities a reciprocal space map was obtained, figure 4.50. From this the integrated areas and relative peak heights could be obtained, compared with the calculated intensities (based on kinematical theory) and the ratio of the two contributions obtained. In this particular case there are nearly equal proportions of the two twin components.

4.7.3.2. Analysing twin components larger than 5 microns

The analysis of features greater than 5 microns can become a problem for diffractometry since the influence on the scattering is not very different from that of bulk samples. However there are orientation and strain effects that separate these features from the surrounding matrix and these can be observed in diffraction, although judging their dimensions is best done by topography. It is interesting to note that the topography and diffractometry procedures for determining size of features are complementary and cover the full range of dimensions. Topography can be applied very successfully in combination with diffractometry and examples have been given in previous sections in this chapter.

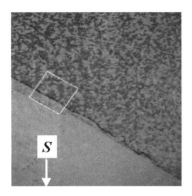 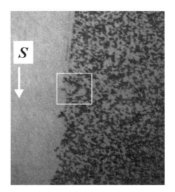

Figure 4.51. The evidence of twins in CdHgTe grown on CdTe (111). The two Berg-Barrett topographs were compared after a 60^0 rotation showing the reversal of contrast. The exposure times were a few minutes and the twin dimensions were approximately 20µm.

This particular example is rather interesting in that a topograph from planes parallel to the surface will yield an even distribution of intensity. The sample is $Cd_{0.2}Hg_{0.8}Te$ grown on a (111) CdTe substrate. A Laue photograph indicates 6-fold symmetry, which in itself is suspicious, it should be 3-fold. This is the first indication of twinning but gives no real indication of size of these twins. A simple Berg-Barrett topograph, or any reflection topography system, e.g. figure 3.27, where a photographic emulsion is held immediately after the sample, will yield images as in figure 4.51. The two images are both of the 115 type reflections, the second after rotation of the sample about the surface normal through 60^0. The images are negatives of each other and clearly indicate that the two contributions are interlocking 180^0 twins. The exposure times are only a few minutes and clearly give a very quick analysis of this type of materials problem.

4.8. Analysis of textured polycrystalline semiconductors

Working through these various examples we have ordered them in decreasing levels of "quality" from nearly perfect epitaxy, through partially relaxed structures to textured epitaxy. When the orientation dependence in the plane of the interface breaks down altogether the sample is essentially textured polycrystalline. When the scattering from planes parallel to the surface are predominately strong whereas those from planes inclined to the

276   X-RAY SCATTERING FROM SEMICONDUCTORS

surface are weaker than expected the sample may well be of textured polycrystalline form. The GaN sample structure on a sapphire substrate described in section 4.7.2 has this characteristic and to confirm this possibility the scattering from planes normal to the surface plane were studied.

The experimental set-up is described in section 3.7.3.2 and figure 3.23. Since this is slit based and low-angular resolution diffractometry the intensity is high and a full range of scattering peaks were found using this in-plane scattering geometry, figure 4.52. If the material was nearly perfect epitaxy then only a few reflections should be observed, figure 4.53. From figure 4.52 the large range of reflections clearly indicates that the structure is composed of a large range of orientated crystallites with a very strong *00l* texture normal to the surface. This information including the simulation of the reciprocal space map now gives a very complete picture of this sample.

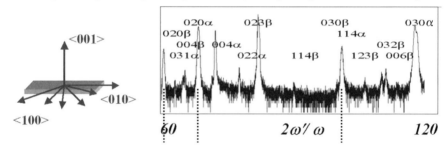

Figure 4.52. A low-resolution in-plane scan from a poor quality epitaxial GaN layer. The large number of reflections is indicative of many crystallites of different orientations.

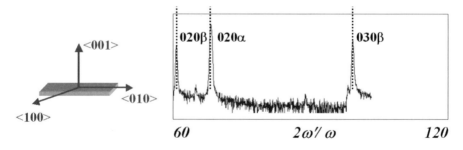

Figure 4.53. A low-resolution reciprocal space map of a good quality GaN layer indicating the expected 001, 002 and 003 reflections. This should be compared with figure 4.52.

4.9. Analysis of nearly perfect polycrystalline materials

As the degree of texture declines and the crystallites in the sample become more randomly orientated the information that characterises the sample becomes less, Table 1.3. The degree of crystallinity can be determined from comparing the crystalline and amorphous peaks, the latter being very broad and weak requiring careful data collection. If there is a distribution of crystallite sizes then this can be obvious from the peak shape (equation 2.94), i.e. less extreme than the amorphous / crystalline comparison, figure 4.54. For determining thicknesses of polycrystalline or amorphous layers then reflectometry is the most appropriate method. From section 2.8 we indicated how the shape of the reflectometry profile is influenced by the density of the material, which in turn relates to the composition (although this can be modified by the porosity or compactness of the crystallites).

Thin polycrystalline layers are important as conducting layers for large area electronics and used in magnetic storage devices, etc., and their characterisation in terms of thickness, state of strain and composition are often wanted parameters. In this section there are brief examples of the various characterisation methods.

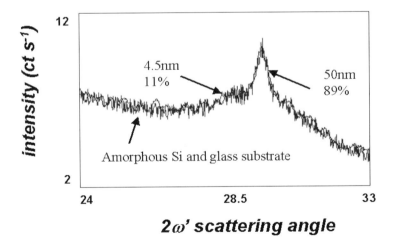

Figure 4.54. The scattering from small crystallites on glass indicating the two distinct size contributions to the scattering. The profiles have been fitted by overlapping Gaussian profiles with noise to aid the comparison.

4.9.1. Measurement of thickness of CrO_x on glass

The CrO_x layer is a textured polycrystalline layer on an amorphous substrate. This is an ideal case for reflectometry provided the sample is not excessively bent, although the effect can be minimised by restricting the beam size on the sample with a knife-edge and / or using multiple crystal methods, sections 3.8.3 and 3.27. These improve the shape of the critical edge.

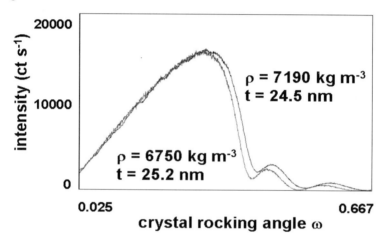

Figure 4.55. The sensitivity of the critical edge of a reflectometry profile to differences in density. The data was collected with the multiple-crystal diffractometer to establish an absolute angular scale and hence absolute density.

The method that gives the most perfect profile is that using multiple crystal methods (with a line focus) because the instrument function is insignificant and all the contributions to the shape are related to the sample. This is the best configuration for complicated multi-layers requiring modelling of the intensity or those that contain thick layers (>300nm), when high angular resolution is required. This will allow the density to be determined to within <1%. However the intensity is compromised and generally the incident beam is more extended and therefore the signal to noise can be a problem at larger scattering angles. The addition of an X-ray mirror can boost the intensity by an order of magnitude, but at the expense of increasing the beam size on the sample, consequently the intensity gain at low angles of incident can be minimal.

The geometry used in this example is as described above (without the X-ray mirror) to place the profile on an absolute angular scale, because the density (that is related to the critical angle) was the important parameter to indicate the oxidisation state. The higher the oxygen concentration the lower the density. Because the scattering angle can easily be placed on an absolute scale, section 4.3.4, the position of the whole profile can be placed on an absolute angle scale. Because of sample size effects and the relatively large incident beam size the shape of the critical edge cannot be defined too closely and the whole shape has to be modelled. Since all the parameters except the thickness and density are known this can be fitted. Figure 4.55 gives an indication of the sensitivity of this approach. The density of pure Cr is 7190kg m^{-3} and if we assume that O substitutes the Cr directly, with minimal lattice parameter change then the oxygen state gives $x = 0.09$.

To boost the intensity simple slit optics can be used. The optimum configuration is with an incident beam divergence of $1/32^0$ or $1/64^0$ and a scattered beam slit of 0.1mm and a knife-edge close to the sample, section 3.7.1.1. Again the profile is distorted due to sample size effects and this should be accounted for in simulating the shape. The dynamic range with this arrangement can be significantly more than the detector can accommodate, therefore filters should be used either automatically or performing several runs and scaling the scans to a common level. With this arrangement the density can only be measured reliably to within about 5%.

To maintain very high intensity, to prevent the need for filters and keep a constant sampled area, one very successful method is to use automatic divergence and scatter slits, section 3.7.4.1. At low angles of incidence the intensity is reduced by restricting the incident beam divergence, which is gradually increased as the incident angle increases. Therefore as the specular beam weakens the incident beam intensity increases. The intensity has to be scaled to relate it to that given by other methods. Generally curvature is a problem in reflectometry since the beam will extend over large areas of the sample. This is a particular problem with fixed slits since the region of the sample analysed varies with angle. Now since the divergence arriving at any point is the same, providing the focus is very small or crystal optics are used to give a parallel beam, the region that satisfies the scattering condition is unchanged yet the area analysed decreases with incident angle. Therefore the proportion of the incident beam that satisfies the scattering condition is increased. This can lead to an

280   X-RAY SCATTERING FROM SEMICONDUCTORS

underestimate of the rate of fall in intensity and consequently overestimate the roughness of the interfaces, etc. The use of variable slits will overcome this problem by maintaining a fixed area of illumination. To reduce or separate the diffuse scattering from the specular contribution careful matching of the scattered beam analyser to the incident beam divergence is very important.

4.9.2. Analysis of very weak scattering

Randomly orientated polycrystalline materials as thin layers will have very weak scattering and at times the scattering of interest is difficult to observe. With the help of reciprocal space mapping and a low angular resolution diffractometer (section 3.7.2 and 3.7.3.2) exceedingly weak scattering can be observed. This can be important for identifying phases, determining the crystallite size or for analysing the state of strain in the plane of the interface for example. We shall just illustrate two examples using these two different geometries to illustrate what is possible with relatively simple apparatus.

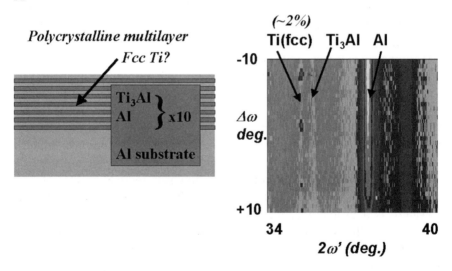

Figure 4.56. The enhancement of exceedingly weak scattering by applying low-resolution reciprocal space mapping.

The first example is a method to determine whether Ti exists in the FCC form in a relatively hard but light Ti / Al multi-layer. This was detected by TEM and was considered an artefact of the TEM sample preparation and hence using X-ray methods with no sample preparation should resolve this issue. The geometry of the apparatus was as described in section 3.7.2, essentially the Bragg-Brentano geometry. A very careful radial scan with $\omega = 2\omega'/2$ indicated no Ti FCC present, however when a limited area low resolution reciprocal space map using the same geometry was carried out the scattering associated with Ti FCC is clearly seen, figure 4.56. The difficulty in seeing this sort of detail using conventional methods also becomes rather obvious, the composition of this phase is very low (a few %) and also that which does exist indicates orientation texture and therefore finding this with a single scan would be pure chance.

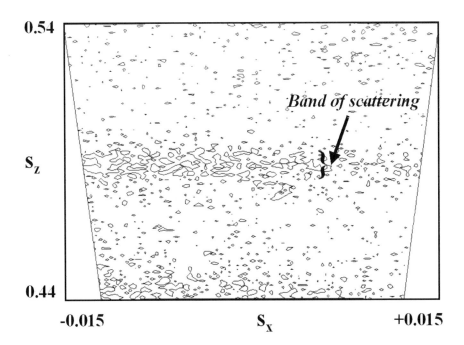

Figure 4.57. The application of low-resolution reciprocal space mapping for measuring the in-plane scattering from a 30nm polycrystalline layer. The lattice parameter was determined as 0.28893nm and the width of the profile gives a crystallite size of 12nm (110 reflection CuKα radiation).

The second example uses the in-plane geometry with double pinhole and parallel plate collimator, section 3.7.3.2. The sample is a 30nm randomly orientated polycrystalline Cr layer on glass and this example gives an indication how the very weak scattering from planes normal to the surface plane can be observed by low-resolution reciprocal space mapping. A single $2\omega'$ scan indicates no scattering observable, however the reciprocal space produces a clear band of scattering, figure 4.57. Both these examples are using some of the simplest optics and a sealed 2kW X-ray Cu source.

4.10. Concluding remarks

These examples should indicate the sensitivity to various parameters and the level of detail that can be achieved. X-ray scattering reveals deviations from perfection with the right method, and very quick analyses are achievable for all the basic parameters. The study of imperfect epitaxy offers the greatest challenge since more parameters are required to characterise the material, Table 1.3, chapter 1. Examples of all these have been covered. There are still many challenging problems to be addressed as the diversity of materials and combinations keep appearing. This subject is still very much alive and dynamic!

References

Auvray, P, Baudet, M and Regreny, A (1989) J Cryst. Growth **95** 288
Birch, J, Sundgren, J-E and Fewster, P F (1995) J Appl. Phys. **78** 6582
Bond, W L (1960) Acta Cryst. **13** 814
Cullity, B D (1978) *Elements of X-ray diffraction,* Addison Wesley, Reading
Dane, A D, Veldhuis, A, de Boer, D K G, Leenaers, A J G and Buydens, L M C (1998) Physica B **253** 254
Dismukes, J P, Ekstrom, L and Paff, R J (1964) J Phys. Chem. **68** 3021
Fewster, P F (1984) J Appl. Cryst. **17** 265
Fewster, P F (1987) *Thin Film Growth Techniques for Low Dimensional Structures* NATO ASI Series B: Physics **163** pp417-440, Ed: Farrow et al. New York: Plenum.
Fewster, P F (1988) J Appl. Cryst. **21** 524

Fewster, P F (1990) Proceedings Volume **90-15**, pp.381-384, J Electrochem. Soc. *Superlattice Structures and Devices*.
Fewster, P F (1986) Philips J Res. **41** 268
Fewster, P F (1991a) *Analysis of Microelectronic Materials and* Devices Chapter 3.1.1. Ed: Grausserbauer and Werner, Wiley: New York
Fewster, P F (1991b) J Appl. Cryst. **24** 178
Fewster, P F (1993) Patent Nos: EPO 0603943B, 5442676, B33826
Fewster, P F (1993) Semicond. Sci. Technol. **8** 1915
Fewster, P F (1996) Critical Reviews in Solid State and Materials Sciences **22** 69
Fewster, P F (1996) Rep. Prog. Phys. **59** 1339
Fewster, P F (1998) *X-ray and Neutron Dynamical Diffraction: Theory and Applications*, pp289-299, Ed: Authier et al, Plenum Press: New York.
Fewster, P F and Andrew, N L (1993a) J Appl. Cryst. **26** 812
Fewster, P F and Andrew, N L (1993b) Materials Science Forum, **133** 221 Ed: Mittemeijer and Delhez, Trans Tech Pub: Switzerland
Fewster, P F and Andrew, N L (1993c) J Appl. Phys. **74** 3121
Fewster, P F and Andrew, N L (1995a) J Phys. D **28** A97
Fewster, P F and Andrew, N L (1995b) J Appl. Cryst. **28** 451
Fewster, P F and Andrew, N L (1998) Thin Solid Films **319** 1
Fewster, P F, Andrew, N L and Curling, C J (1991) Semicond. Sci. Technol. **6** 5
Fewster, P F and Curling, C J (1987) J Appl. Phys. **62** 4154
Fewster, P F and Tye, G A (2001) Patent no: GB010114
Fewster, P F and Whiffin, P A C (1983) J Appl. Phys. **54** 4668
Fullerton, E E, Schuller, I K, Vanderstaeten, H and Bruynseraede, Y (1992) Phys. Rev. **B45** 9292.
Hornstra, J and Bartels, W J (1978) J Cryst. Growth **44** 513
Kaganer, V M, Koehler, R, Schnidbauer, M, Opitz, R and Jenichen, B (1997) Phys. Rev. B**55** 1793
Keissing, H (1931) Ann. Phys. Lpz. **10** 769
Kidd, P and Fewster, P F (1994) Mat. Res. Soc. Symp. Proc. **317** 291
Klappe, J G E and Fewster, P F (1993) J Appl. Cryst. **27** 103
Laugier, J and Filhol, A (1983) J Appl. Cryst. **16** 281
Nagai, H (1974) J Appl.Phys. **45** 3789
Sevidori, M, Cembali, F and Milita, S (1996) *X-ray and Neutron Dynamical Diffraction: Theory and Applications*, pp.301-321, Ed: Authier et al, Plenum Press: New York.

Schiller, C (1988) Analysis **16** 402
Tye, G A and Fewster, P F (2000) Unpublished work
Wormington, M, Panaccione, C, Matney, K M and Bowen, D K (1999) Phil. Trans. Roy. Soc. **A 357** No. 1761 p2827

APPENDIX 1

GENERAL CRYSTALLOGRAPHIC RELATIONS

A.1. Introduction

The crystal symmetry of any material is reflected in its physical properties. The sensitivity of certain crystallographic planes to certain properties therefore can determine the experiment to consider. To help choose a suitable scattering plane for an experiment some commonly used stereographic projections are given for the cubic and hexagonal symmetries. The common surface orientations are *(001), (110)* and *(111)* in the cubic case and *(0001)* in the hexagonal case. The projections for cubic symmetries are very general, however hexagonal projections are altered by the ratio of the *c* to *a* axis lengths. Therefore only the one projection is given for the hexagonal case, although this projection is not too sensitive to the *c/a* ratio. The additional index in the hexagonal notation *(hkil)* can be derived from $i = -\{h + k\}$.

The interplanar spacings for various sets of planes is given for the general case which have then be simplified for the cubic and hexagonal cases. The unit cell lengths are *a*, *b* and *c* and the angles between *a* and *b* is γ, *b* and *c* is α and *c* and *a* is β.

A.2. Interplanar spacings

The distance *d* between successive atomic planes of the type *hkl* is given by

$$d = \left[\frac{1}{V^2}\{s_{11}h^2 + s_{22}k^2 + s_{33}l^2 + 2s_{12}hk + 2s_{13}hl + 2s_{23}kl\}\right]^{-1/2}$$

A1

where

$$s_{11} = \{bc\sin(\alpha)\}^2$$
$$s_{22} = \{ca\sin(\beta)\}^2$$
$$s_{33} = \{ab\sin(\gamma)\}^2$$
$$s_{12} = abc^2\{\cos(\alpha)\cos(\beta) - \cos(\gamma)\}^2$$
$$s_{13} = ab^2c\{\cos(\gamma)\cos(\alpha) - \cos(\beta)\}^2$$
$$s_{23} = a^2bc\{\cos(\beta)\cos(\gamma) - \cos(\alpha)\}^2$$

A.2

and the volume is given by

$$V = abc\{1 - \cos^2(\alpha) - \cos^2(\beta) - \cos^2(\gamma) + 2\cos(\alpha)\cos(\beta)\cos(\gamma)\}$$

A.3

We can simplify these relationships for the cubic symmetry

Cubic:

$a = b = c$
$\alpha = \beta = \gamma = 90°$

hence

$$d = \frac{a}{\left[h^2 + k^2 + l^2\right]^{1/2}}$$

A.4

$$V = a^3$$

A.5

Hexagonal:

$a = b \neq c$

$\alpha = \beta = 90^0, \gamma = 120^0$

$$d = \left[\frac{3}{4 \left\{ \frac{h^2 + k^2 + hk}{a^2} + \frac{3l^2}{4c^2} \right\}} \right]^{1/2} \qquad \text{A.6}$$

$$V = \frac{3^{1/2} a^2 c}{2} \qquad \text{A.7}$$

A.3. Stereographic projections:

The atomic plane normals are denoted by the indicies hkl and from the origin they will intersect a hemisphere at the locations given in figure A1 to A4. Thus for the *(001)* projection the *(001)* plane normal will be directly above the origin. The distances along the surface of this sphere will represent the angles between any two planes. For example the angle between *(001)* and *(100)* is 90^0 in the cubic system and will therefore will occur at a distance of one quarter of the circumference of the sphere. The angles between common directions in the cubic system are given in Table A1. The lines in the projection are a guide to follow surface normals of a type, for example in figure A1 the line from *(001)* to *(110)* will include *(111)*, *(112)*, *(113)*, etc., remembering also that these are parallel to *(333)*, *(224)* and *(226)* respectively.

The angles between atomic planes $(h_1k_1l_1)$ and $(h_2k_2l_2)$ are determined from the following formulae.

The angle between atomic planes is given by

$$\varphi = \cos^{-1}\left(\frac{d_1 d_2}{V^2} \left\{ s_{11} h_1 h_2 + s_{22} k_1 k_2 + s_{33} l_1 l_2 + s_{12}(h_1 k_2 + h_2 k_1) + s_{13}(h_1 l_2 + h_2 l_1) + s_{23}(k_1 l_2 + k_2 l_1) \right\} \right)$$

A.8

Again this expression can be simplified for cubic and hexagonal systems

Cubic:

$$\varphi = \cos^{-1}\left(\frac{h_1 h_2 + k_1 k_2 + l_1 l_2}{\left[(h_1^2 + k_1^2 + l_1^2)(h_2^2 + k_2^2 + l_2^2) \right]^{1/2}} \right)$$

A.9

Hexagonal:

$$\varphi = \cos^{-1}\left(\frac{h_1 h_2 + k_1 k_2 + \frac{1}{2}(h_1 k_2 + h_2 k_1) + \frac{3a^2}{4c^2} l_1 l_2}{\left[\left(h_1^2 + k_1^2 + h_1 k_1 + \frac{3a^2}{4c^2} l_1^2 \right)\left(h_2^2 + k_2^2 + h_2 k_2 + \frac{3a^2}{4c^2} l_2^2 \right) \right]^{1/2}} \right)$$

A.10

etc.

Appendix 1 General Crystallographic Relations 289

Table A.1: The angles between common crystallographic planes in the cubic system:

(hkl)	Common reflections for FCC semiconductors	1 0 0	1 1 0	1 1 1
1 0 0	002, 004, 006	0 90	45 90	54.7
1 1 0	022, 044	45 90	0 60 90	35.3 90
1 1 1	111, 222, 333, 444	54.7	35.3 90	0 70.5 109.5
2 1 0	024	26.6 63.4 90	18.4 50.8 71.6	39.2 75
2 1 1	224	35.3 65.9	30 54.7 73.2 90	19.5 61.9 90
2 2 1	244	48.2 70.5	19.5 45 76.4 90	19.5 61.9 90
3 1 0	026	18.4 71.6 90	26.6 47.9 63.4 77.1	43.1 68.6
3 1 1	311	25.2 72.5	31.5 64.8 90	29.5 58.5 80
3 2 0	046	33.7 56.3 90	31.5 64.8 90	29.5 58.5 80
3 2 1	246	36.7 57.7 74.6	19.1 40.9 55.5	22.2 51.9 72 90
3 3 1	331	46.5 76.7	13.1 49.5 71.1	22 48.5 82.4
3 3 5	335	62.8	49.7	14.4 84.9
5 1 1	115	15.6 78.9	74.2	38.9
7 1 1	117	11.3 82	78.6	43.3

290 X-RAY SCATTERING FROM SEMICONDUCTORS

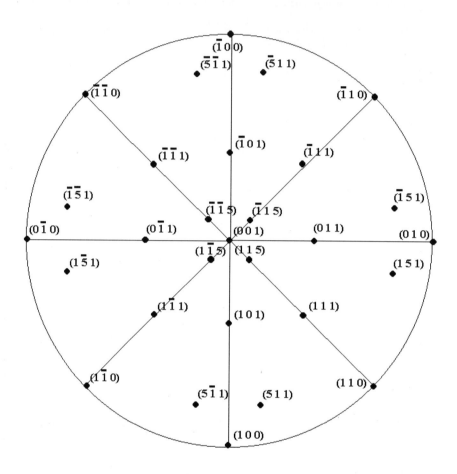

Figure A.1. Stereographic projection for the cubic system *(001)* orientation.

Appendix 1 General Crystallographic Relations 291

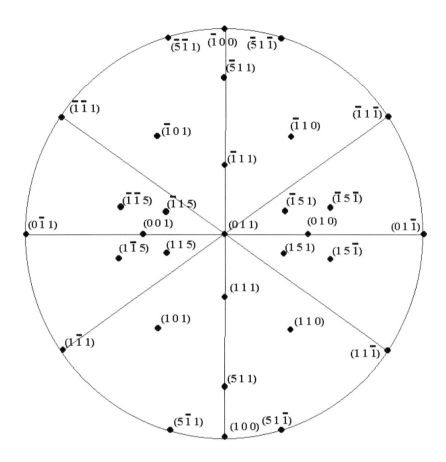

Figure A.2. Stereographic projection for the cubic system *(011)* orientation.

292 X-RAY SCATTERING FROM SEMICONDUCTORS

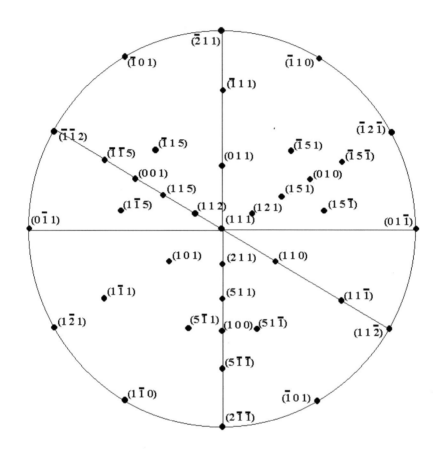

Figure A.3. Stereographic projection for the cubic system *(111)* orientation.

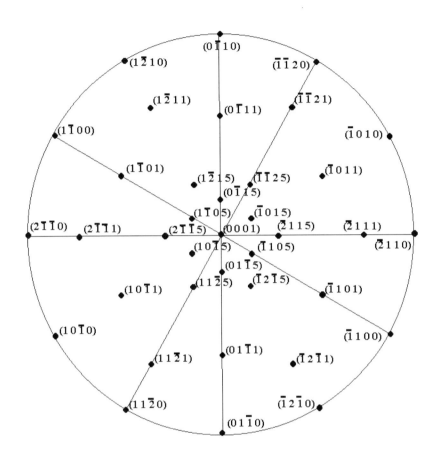

Figure A.4. Stereographic projection for the hexagonal system *(0001)* orientation for a *c/a* ratio of 1.63 (similar to that for GaN).

SUBJECT INDEX

A

Absorption 10, 24, 36, 41, 42, 45, 48, 68, 69, 73, 74, 83, 84, 89, 96, 100, 101, 102, 114, 115, 117, 123, 128, 180, 182, 185, 197, 234, 235, 240
Alignment 152, 161
Amorphous materials 2, 4, 5, 6, 13, 21, 61, 63, 88, 277, 278
Amplitude ratio 50, 58, 59, 60, 62, 63
Anomalous transmission, Borrmann effect 102
Auger process 25, 115
Automatic divergence slits 124, 150
Automatic fitting of profiles 215, 245
Axial divergence 82, 110, 111, 121, 124, 126, 127, 135, 136, 138, 147, 150, 157, 164, 165, 167, 168, 189, 191, 196, 248

B

Beam expansion, compression 53
Borrmann effect, Anomalous transmission 102
Borrmann triangle 100
Boundary conditions 41, 64, 65, 66, 97, 99
Bragg-Brentano diffractometer ... 141, 142, 145, 146, 150, 281
Bulk semiconductors 173

C

Capillary lens 140
Channel-cut or grooved crystals .. 130
Coherence 10, 26, 62, 67, 103, 225
Coherent scattering 7, 24, 25, 28, 30, 62, 72, 110, 141, 145, 234, 240
Composition 1, 3, 4, 5, 13, 17, 79, 105, 145, 171, 172, 173, 195, 200, 203, 204, 205, 208, 209, 210, 212, 214, 218, 224, 227, 228, 229, 230, 231, 234, 235, 236, 237, 238, 240, 241, 242, 243, 246, 249, 250, 251, 251, 252, 253, 254, 255, 256, 258, 260, 262, 264, 271, 277, 281
Composition determination 200
Compton scattering 23, 117, 119, 137, 146
Critical angle 89, 99, 139, 148, 179, 279
Critical thickness 20, 172
Curvature measurement 197

D

Darwin theory of scattering ... 34, 103
Data collection methods 66, 105, 124, 152, 154, 159, 160, 162, 164, 166, 167, 169, 210, 219, 233, 241, 256, 277
Debye-Waller factor, temperature factor 31, 32, 63, 80
Definition of a crystal 12
Degree of relaxation 18, 222, 255, 257, 260, 261, 265

295

Deviation parameter 50, 52, 53, 54, 59
Dielectric constant 35
Diffuse scattering 6, 77, 81, 82, 91, 92, 94, 95, 131, 141, 143, 160, 166, 168, 192, 193, 194, 219, 220, 221, 222, 226, 227, 239, 247, 250, 264, 265, 280
Diffuse scattering topograph 194
Direction cosines 50, 51, 52, 53
Dislocation density measurement 223, 263
Dispersion surface 44, 45, 47, 49, 53, 54, 55, 67, 83, 99, 101, 108
Displacement field 39, 41, 43, 56, 59, 99
Distorted Born Wave Approximation, DWBA 92
Distorted interface 62, 63
Distorted wave vector 59
Divergence in the scattering plane 110, 121, 126, 165, 167
Double crystal diffractometer 151, 152, 154, 203, 261

E

Elastic parameters 13, 17, 204, 205, 254
Elastic properties 13, 82
Elastic scattering 24, 25, 110
Electron density 33, 35, 37, 38, 83, 88, 90
Electron radius 56
Energy discrimination 115, 116, 118
Energy flow in a crystal 47, 101, 102
Energy resolution, detector 116, 117, 118, 120, 145
Escape peak 115, 116
Ewald theory of dynamical scattering 34

Excitation points, tie points 46, 47, 48
Extinction 47, 68, 256

F

Fluorescence 25, 117
Form factor 30
Forward refracted wave 57, 68
Fractal interface roughness 92

G

Gallium arsenide 2, 3, 12, 15, 17, 20, 76, 154, 168, 174, 175, 180, 182, 185, 186, 187, 190, 191, 193, 200, 202, 208, 213, 225, 230, 235, 241, 242, 243, 244, 245, 246, 247, 248, 250, 257, 258, 260, 263, 264, 265
Gallium nitride 12, 18, 19, 270, 271, 272, 276, 293
Gas amplification 115, 116, 117, 120
Growth mechanisms 19

I

Incident beam filters 128
Incident beam monochromator 146
Incoherent scattering 25
Incommensurate structures 232, 240
In-plane scattering 96, 97, 147, 149, 167, 168, 276, 281
Instrument function 7
Interfaces 5, 61, 89, 90, 93, 99, 210, 226, 227, 229, 240, 242, 246, 248, 264, 265, 280
Interplanar angles 287
Interplanar spacings 285
Ion-implanted materials 210

K

Kato spherical wave theory 102

Kinematical theory of scattering...68, 240, 245
Knife-edge 143, 144, 198, 199, 278, 279

L

Laboratory sources............. 9, 10, 111
Lang method 184, 185, 186
Lateral correlation length........ 74, 92, 93, 266, 267, 268, 269, 270, 271
Lateral interface roughness............ 91
Laterally periodic structures 76
Lattice parameter determination .. 195
Laue method 174, 182
Laue's equations 27, 28, 38
Laue theory of dynamical scattering.................................35
Layer linking............................... 213
Layer tilt 5, 17, 106, 149, 153, 154, 160, 163, 164, 178, 179, 182, 183, 189, 192, 196, 197, 200, 201, 211, 219, 223, 224, 225, 226, 248, 250, 256, 261, 264, 265, 266, 267, 268, 269, 270
Linear absorption coefficient........ 73, 84, 100, 101, 102
Lithium niobate.................... 182, 183
Low-resolution reciprocal space mapping............................ 280, 281

M

Macroscopic properties........ 5, 6, 7, 8, 56, 57, 83, 88
Mass absorption coefficient............ 73
Microscopic properties 5, 6, 7, 8, 266, 267, 268, 269, 270
Microscopic tilt........... 266, 267, 268, 269, 270
Miller indices 27, 252
Misfit dislocations 12, 20, 60, 264
Mismatch 255

Mosaic crystals 6, 12, 20, 66, 68, 74, 75, 76, 142, 145, 147, 160, 164, 165, 173, 181, 182, 183, 184, 186, 187, 188, 189, 190, 191, 192, 193, 197, 248, 264, 265, 266, 272
Mosaic structure simulation..... 63, 66
Multiple crystal diffractometer 43, 160, 161, 177, 179, 195, 197, 245, 247, 256, 258, 259

N

Non-coplanar scattering.......... 96, 97, 147, 149, 167, 168, 276, 281

O

Orientation.................... 4, 5, 6, 7, 13, 18, 21, 26, 68, 142, 147, 150, 163, 173, 174, 176, 177, 178, 182, 187, 190, 192, 200, 204, 252, 273, 274, 275, 281, 290, 291, 292, 293

P

Parallel plate collimator.............. 126, 127, 146, 147, 148, 149, 282
Period variation 137, 233, 237, 240, 243
Periodic multi-layers 200, 229, 235, 239, 246, 247
Phase averaging 26, 62, 79
Phonons 78, 80, 81
Photoelectric absorption 25, 34, 68, 117, 256
Photomultiplier 117
Plane wave theory............ 39, 57, 102
Plastic deformation 4, 12, 14, 249
Point defects 5, 12, 77, 131
Poisson ratio 15, 17, 61, 62, 241, 254
Polarisabilit 37, 44, 51, 57
Polarisation................. 10, 35, 42, 43, 46, 58, 86, 101, 185

Polarit 180, 181
Polycrystalline materials 5, 6, 7, 12, 13, 21, 88, 121, 125, 126, 138, 141, 145, 147, 150, 189, 242, 266, 275, 277, 278, 280, 281, 282
Position sensitive detectors .. 119, 120
Proportional detector 114, 116

Q

Quantum dots 168
Quantum well structures 227, 228, 229

R

Radius of curvature determination 197, 198, 199
Rayleigh scattering 24
Relaxation, degree of 5, 13, 18, 20
Reciprocal lattice point 108, 110, 153, 159, 163, 164, 166, 192, 252
Reciprocal space co-ordinates 106, 162
Reciprocal space map simulation ... 64
Reciprocal space mapping 2, 64, 82, 145, 155, 162, 164, 166, 167, 173, 187, 189, 192, 197, 221, 247, 250, 251, 256, 259, 260, 264, 266, 273, 280, 282
Reflectometry 143, 144, 148, 150, 168, 200, 205, 208, 238, 240, 245, 246, 277, 278, 279
Reflectometry scattering theor 82
Refractive index 32, 33, 44, 45, 51, 54, 55, 64, 65, 67, 68, 69, 83, 84, 85, 86, 89, 138, 177, 197, 209, 239
Relaxation, degree of 5, 13, 18, 20, 119, 168, 172, 195, 202, 223, 240, 249, 250, 251, 255, 256, 258, 260, 262, 263, 264, 272
Relaxed structures 226, 249, 256, 275

Resolution in topograph 191
Resolution of a diffractometer 108, 151
Resonant scattering 36, 38
Roughness, interfacial 90
Roughness, vertical correlation 93
Rocking-curve 2, 140, 159, 160, 178, 187, 210, 211, 220, 223, 224, 225, 226, 231, 236, 256, 258, 259, 261, 262

S

Sample inhomogeneit 266
Satellites 80, 81, 136, 232, 233, 235, 238, 240, 242, 243, 245
Scattered beam monochromator .. 145
Scattered wave 29, 32, 39, 43, 45, 47, 50, 54, 57, 60, 64, 65, 67, 68, 82, 97, 98, 101, 102, 132, 180
Scattering factor 30, 31, 36, 38, 77, 180, 242
Scattering vector 27, 29, 31, 33, 39, 47, 49, 54, 64, 79, 82, 87, 97, 108, 109, 158, 197, 248
Scintillation detector 117, 118, 119
Scherrer equation 73, 205
Segregation 90, 227, 228
Semiconductor wafers 173
Silicon 2, 12, 25, 90, 91, 99, 119, 158, 185, 204, 208, 210, 211, 212, 213, 218, 219, 221, 222, 223, 259
Silicon-germanium alloys 210, 212, 213, 218, 221, 259
Simulation of rocking curves 209
Solid state detectors 118, 119, 120
Soller slits 126, 138, 150, 167
Specular reflectometry 43
State of strain 5, 11, 12, 13, 16, 18, 20, 57, 61, 62, 63, 68, 75, 82, 99, 160, 172, 173, 187, 188, 189, 190, 192, 195, 197, 200, 202,

203, 210, 221, 226, 227, 229, 235, 239, 241, 242, 249, 250, 251, 253, 255, 256, 257, 264, 265, 274, 277, 280
Stiffness coefficients 14
Structure factor 38, 69, 70, 78, 88, 180, 231, 232, 234, 235, 238, 240, 242, 244
Superlattices 80, 95, 145, 225, 229, 231, 232, 235, 236, 237, 239, 241, 242, 243, 244, 245, 246, 247, 264, 265, 283
Surface damage 131, 174, 192, 194, 197
Susceptibilit 37, 43, 46, 56, 61, 88, 101
Stereographic projections 285, 287
Synchrotron sources 9, 10, 42, 111

T

Takagi theory 56
Temperature factor 31, 32, 80
Thickness determination 205, 208
Three-dimensional reciprocal space mapping 111, 165, 187, 189, 190, 197, 248
Tie points, excitation points ... 47, 101
Tilt alignment 211, 219
Tilting on vicinal surfaces 201
Topography 7, 10, 20, 59, 99, 100, 102, 119, 166, 167, 173, 182, 183, 184, 185, 186, 187, 190, 191, 192, 222, 226, 263, 264, 266, 273, 274, 275
Toroidal mirrors 138

Total external reflection 45, 139, 209
Transmission scattering geometry 99, 100, 102
Triple crystal diffractometer 154, 155
Twinning measurements 273, 275
Two beam scattering 45
Two-crystal four-reflection monochromator 133, 135, 152, 177, 198, 210, 212, 219

U

Unit cell 7, 12, 16, 18, 27, 37, 38, 57, 88, 180, 200, 230, 231, 233, 240, 253, 281, 285

V

Vegard's rule 204, 205, 253, 254
Vegard's rule deviation 204

W

Wiggles in scattering 193, 225

X

X-ray detectors 23, 113, 167
X-ray diffractometer 105, 141
X-ray mirrors 11, 137, 156, 219, 245, 247, 278, 279
X-ray sources 11, 23, 32, 102, 105, 106, 107, 112, 113, 121, 124, 126, 127, 128, 129, 132, 134, 138, 139, 141, 158, 165, 167, 172, 175, 182, 259
X-ray tube focus size ... 112, 122, 142